Biodiversity and Conservation

Rio 1992. Presidents and princes from around the world signed the United Nations convention on Biodiversity, but less than a decade earlier no one had heard the word biodiversity. What was so special, so urgent that this short word for 'biological diversity' entered the global vocabulary?

Biodiversity is the total richness and variety of life on earth, it is the natural power-house of planetary health, it is a new science born of ecology, evolutionary and conservation biology. It is facing a terrible threat.

This book reviews the conceptual origins of the term biodiversity, the combination of biological insight and environmental awareness that lead to the Rio treaty. This is linked to the epic history of Life On Earth, culminating in the current crisis of extinction and ecosystem collapse. The underlying ecological patterns and processes, plus evolutionary forces, that create biodiversity are outlined. The resulting inventory of life's variety – genetic, taxonomic and ecological – is described in chapter 3. The dangers, natural and human, are discussed in chapter 4; then, in chapter 5, the approaches, old and new, to conserve nature's riches are discussed.

This is a unique springboard from which those approaching the subject for the first time can explore this vital topic, from founding papers to the latest triumphs and tragedies.

Michael J. Jeffries is a Lecturer in Ecology and Environment at the University of Northumbria.

Routledge Introductions to Environment Series
Published and Forthcoming Titles

Titles under Series Editors:
Rita Gardner and Antoinette Mannion

Environmental Science texts

Environmental Biology
Environmental Chemistry and Physics
Environmental Geology
Environmental Engineering
Environmental Archaeology
Atmospheric Systems
Hydrological Systems
Oceanic Systems
Coastal Systems
Fluvial Systems
Soil Systems
Glacial Systems
Ecosystems
Landscape Systems

Titles under Series Editors:
David Pepper and Phil O'Keefe

Environment and Society texts

Environment and Economics
Environment and Politics
Environment and Law
Environment and Philosophy
Environment and Planning
Environment and Social Theory
Environment and Political Theory
Business and Environment

Key Environmental Topics texts

Biodiversity and Conservation
Environmental Hazards
Natural Environmental Change
Environmental Monitoring
Climatic Change
Land Use and Abuse
Water Resources
Pollution
Waste and the Environment
Energy Resources
Agriculture
Wetland Environments

Energy, Society and Environment
Environmental Sustainability
Gender and Environment
Environment and Society
Tourism and Environment
Environmental Management
Environmental Values
Representations of the Environment
Environment and Health
Environmental Movements
History of Environmental Ideas
Environment and Technology
Environment and the City
Case Studies for Environmental Studies

Routledge Introductions to Environment

Biodiversity and Conservation

Michael J. Jeffries

London and New York

First published in 1997
by Routledge
11 New Fetter Lane, London EC4P 4EE

Simultaneously published in the USA and Canada
by Routledge
29 West 35th Street, New York, NY 10001

Typeset in Times by Keystroke, Jacaranda Lodge, Wolverhampton

Printed and bound in Great Britain by Redwood Books, Trowbridge, Wiltshire

British Library Cataloguing in Publication Data
A catalogue record for this book is available from the British Library

Library of Congress Cataloguing in Publication Data
Jeffries, Mike
    Biodiversity and conservation / Mike Jeffries.
        p.   cm. — (Routledge introductions to environment series)
    Includes bibliographical references and index.
    1. Biological diversity. 2. Biological diversity conversation.
    I. Title. II. Series: Routledge introductions to environment.
    QH541.15.B56J44   1997
    333.95—dc21   96-47381

ISBN 0–415–14904–5
        0–415–14905–3 (pbk)

Contents

Series editors' preface
Environmental Science titles

The last few years have witnessed tremendous changes in the syllabi of environmentally related courses at Advanced Level and in tertiary education. Moreover, there have been major alterations in the way degree and diploma courses are organised in colleges and universities. Syllabus changes reflect the increasing interest in environmental issues, their significance in a political context and their increasing relevance in everyday life. Consequently, the 'environment' has become a focus not only in courses traditionally concerned with geography, environmental science and ecology but also in agriculture, economics, politics, law, sociology, chemistry, physics, biology and philosophy. Simultaneously, changes in course organisation have occurred in order to facilitate both generalisation and specialisation; increasing flexibility within and between institutions is encouraging diversification and especially the facilitation of teaching via modularisation. The latter involves the compartmentalisation of information which is presented in short, concentrated courses that, on the one hand are self-contained but which, on the other hand, are related to prerequisite parallel and/or advanced modules.

These innovations in curricula and their organisation have caused teachers, academics and publishers to reappraise the style and content of published works. While many traditionally styled texts dealing with a well-defined discipline, e.g. physical geography or ecology, remain apposite there is a mounting demand for short, concise and specifically focused texts suitable for modular degree/diploma courses. In order to accommodate these needs Routledge has devised the Environment Series which comprises Environmental Science and Environmental Studies. The former broadly encompasses subject matter which pertains to the nature and operation of the environment and the latter concerns the human dimension as a dominant force within, and a recipient of, environmental processes and change. Although this distinction is made, it is purely arbitrary and for practical rather than theoretical purposes; it does not deny the holistic nature of the environment and its all-pervading significance. Indeed, every effort has been made by authors to refer to such interrelationships and provide information to expedite further study.

This series is intended to fire the enthusiasm of students and their teachers/lecturers. Each text is well illustrated and numerous case studies are provided to underpin general theory. Further reading is also furnished to assist those who wish to reinforce and extend their studies. The authors, editors and publishers have made every effort to provide a series of exciting and innovative texts that will not only offer invaluable learning resources and supply a teaching manual but also act as a source of inspiration.

A. M. Mannion and Rita Gardner
1997

Series International Advisory Board

Australasia: Dr Curson and Dr Mitchell, Macquarie University

North America: Professor L. Lewis, Clark University; Professor L. Rubinoff, Trent University

Europe: Professor P. Glasbergen, University of Utrecht; Professor van Dam-Mieras, Open University, The Netherlands

Note on the text

Bold is used in the text to denote words defined in the Glossary. It is also used to denote key terms.

Plates

Figures

Tables

Boxes

Author's preface

September 1992 – and I was marooned up an Acacia tree in the Zimbabwean bush while an elephant meandered below, picking up the tasty pods. The African Elephant is inspiring, gentle, bound by strong family ties, a vital player in the ecology of the bush but victim of human greed. Or, if you prefer, it is the destroyer of crops and homes, a deadly danger killing defenceless farmers every year and a potential source of income to the rural poor of developing Africa denied to them by sentimental, affluent Westerners. As a candidate symbol uniting the themes of biodiversity and conservation the African Elephant is ideal: a species with a fine evolutionary history, central to the ecological workings of its habitat, threatened by human pressures, the focus of intense conservation work and debate. As for me up the tree – this is not quite as heroic as it sounds. The tree had a viewing platform and mini-bar so the main sacrifice was temporary loss of the barman who climbed down to drive the beast away by banging the ice-tongs on a drinks tray. In addition to its symbolic and ecological roles the elephant was also providing income to locals, via my bar bill. The bar was stocked from the fruits of the ancient metabolic skills of microbes, producing alcohol and reached by aeroplane and car, powered by fossil fuels distilled from the remains of ancient ecosystems – in the process creating environmental destruction directly through exhaust fumes, indirectly at the refining stage. It is impossible to do anything without feeling the gravitational pull of biodiversity on our lives.

I hope this book captures something of the mystery, excitement and possibilities that attend the new science of biodiversity. If so it will be little thanks to my expertise in the field. The book is a shameless pillaging of other people's insights, knowledge and energy. If there is any consolation for the original thinkers whose work I have ransacked I hope it is that the inclusion of their ideas, names and references for further reading reflect the inspiration I have received. Students should read their work. They should dive into the topic using this book as a springboard.

I am very grateful to those who have helped with its production: Phil O'Keefe who sent me to Africa; the editors and reviewers of early drafts; Chris Gibbins for help with photographs; Gary Haley for the illustrations; Penny Allport for her copy editing; Anne Owen, Casey Mein and Sarah Lloyd at Routledge who I suspect wish I was still up that tree.

Michael Jeffries

Introduction: Rio

June 1992. Representatives from over 180 governments met in Rio de Janeiro for the Earth Summit, the second United Nations Conference on Environment and Development.

Twenty years earlier the first UN conference on the Human Environment met in Stockholm. Widely pilloried as a dialogue of the deaf, the Stockholm conference fractured as developed countries asked poorer nations to clean up environmentally destructive development while the Third World wished for economic development, even if pollution and degradation were the price. The 1992 Rio Earth Summit risked a similar schism with the developed world focused on climate change, destruction of tropical forests and species loss, and the developing world still desperate for economic improvement. But two themes had emerged since Stockholm that encouraged global empathy. First, in 1982 the Brundtland Report provided a defining moment, partly in response to the Stockholm fall out, promoting the concept of **sustainable development**. Second the word **biodiversity**, scarcely heard of a decade before Rio, had gained a global audience.

The Earth Summit spawned four major agreements: the Rio Declaration on Environment and Development (citing the rights and responsibilities of individual states); the Convention on Climate Change; Agenda 21 (wide-ranging objectives and approaches for sustainable development); the Convention on Biological Diversity. This Convention was signed by 155 states at the Summit between 5 and 14 June 1992. Of the 105 princes, presidents and prime ministers who had come to Rio and gave formal speeches (despite much touted threats from gangs of muggers and transvestite carnivals) 67 specifically mention biodiversity (or biological diversity). Biodiversity had become a dominating theme, most famously immortalised by the then US President George Bush. Faced with impending presidential elections Bush would not commit himself to attend, unwilling, as he put it, to save squirrels if it cost one American job. He did attend Rio, did not sign the Convention while he was there and did not get re-elected. However, his successful rivals had maintained a strong environmental agenda. The environment was a political issue.

A global convention, a havering US president and 'biological diversity' as a popular buzzword for 67 heads of state – biodiversity had gained a familiarity greater than any other ecological concept and in a remarkably short space of time. This spectacular debut suggests something more than scientific rigour and academic interest. Biodiversity had caught a wider mood.

1 Biodiversity: From concept to crisis

Biodiversity is a recent concept, life on Earth a very ancient phenomenon. The history of both have fuelled awareness of a global environmental crisis. This chapter covers:

- **Origins of the term biodiversity**
- **Themes embraced by concept of biodiversity**
- **History of life on Earth**
- **Current biodiversity crisis**

The word **biodiversity** was coined in the mid-1980s to capture the essence of research into the variety and richness of life on Earth. The word is now widely used, its rapid establishment in science and popular culture an indication of the importance of the topic but also a source of confusion. First, biodiversity is the richness and variety of life on Earth. The flowers and insects and bacteria and forests and coral reefs are biodiversity. Second, biodiversity is an area of scientific research, including both description and measures of diversity and explanations of how this diversity is created. Biodiversity has been increasingly used as a conceptual focus for conservation policy and practice in response to one of the strongest themes underpinning the founding work on biological diversity, species extinction and ecosystem loss, brought to global prominence by the Rio summit. The variety of life on Earth can be investigated at different levels: genetic variation, the numbers of species, the extent of ecosystems. Nature's creative and destructive forces include ecological and evolutionary processes. Current degradation of biodiversity is driven by human pressures and conservation responds with protective laws, reserves and refuges in captivity. The purpose of this book is to provide an introduction to all these topics. Chapter 1 outlines the history of biodiversity, both as a concept and the story of life on Earth up to the current crisis. Chapter 2 describes the ecological and evolutionary processes that spawn this diversity and govern its distribution. Chapter 3 provides an inventory of how biodiversity can be classified and quantified. Chapter 4 describes recent extinctions and their causes, natural and human. Chapter 5 outlines policy and practice to conserve biodiversity.

First then, how did a word which was unheard of in 1980 come to be at the heart of a United Nations global treaty only twelve years later?

The road to Rio. The conceptual history of biodiversity

Origins

The diversity of life on Earth has been a central theme of the natural sciences but it touches on many other areas. The Bible credits Adam with the job of naming the animals, a fundamental task for those quantifying biodiversity. The same approaches to classifying life are apparent in ancient and modern societies. Western culture has repeatedly revised its understanding of the variety and nature of life. The Greek philosopher Aristotle recognised 500 to 600 species, echoing modern folk classifications which typically recognise 300 to 600. Slavish copying of classical texts was abandoned during the sixteenth and seventeenth centuries, spurred by technological advances and the spread of ideas through printing. Classification increasingly focused on the species. The nineteenth century saw the final abandonment of the folk biology principle that lumped species together by broad type (e.g. tree) in favour of biodiversity described by detailed structure and relatedness. Our attitudes to life continue to change. The twentieth century has been dominated by our understanding of evolution which is so powerful a theory that, according to Richard Dawkins, it has rendered God superfluous. At the same time ancient patterns linger. The common use of 'r' and 'l' sounds in words for 'frog' suggest an intuitive, onomatopoeic appreciation of how we describe living things. The richness of life has been central to human society and science but the term biodiversity is an upstart.

The origin of the term is credited to two papers published in 1980 (Lovejoy 1980; Norse and McManus 1980). Lovejoy, working for the World Wildlife Fund in Washington DC, contributed to the Global 2000 Report to the US President which reviewed global environmental topics such as energy, human population and economics. President Carter's preface identified 'the urgency of international efforts to protect our common environment'. Examining the extent of global forestry resources, including projections for future use, two consequences of forest exploitation (changes to global climate and biological diversity) were reviewed. Estimates of extinctions based on different forest loss rates were given. Lovejoy wrote of biological or biotic diversity defined as the total number of species. He wrote of biological capital, the inability of markets to value properly ecological systems and the functions that ecosystems fulfil that are of benefit to people.

Norse and McManus were ecologists on the White House Council on Environmental Quality during the Carter presidency and contributed a chapter to the Eleventh Annual Report of the Council on Environmental Quality. The chapter examines global biodiversity which is defined as two related concepts, genetic and ecological diversity (the latter equated with numbers of species). The bulk of the chapter discusses the material benefits of biological diversity, psychological and philosophical bases for preservation, human impacts that were losses and strategies and policy for conservation.

The context of these papers is important. Biodiversity was discussed on a global scale. The bulk of the work dealt with wider themes than the purely biological.

The importance of biodiversity, actual and potential, was apparent. There is a recognition that the activity of natural ecosystems provides what are now called services or functions vital to a healthy planet. There is a political and philosophical slant to some of the discussion. This usage carries much of the resonance familiar today, combining the richness of life, an awareness of loss, the importance of biodiversity for economics and an ethical, social dimension. Biodiversity is not solely a branch of biology.

The snappy abbreviation biodiversity (fleetingly BioDiversity) is credited to Walter Rosen, working for the American Natural Research Council/ National Academy of Sciences, as a co-director for the 1986 Conference 'The National Forum on BioDiversity', held in Washington DC. The publication of papers from the Forum in a book entitled *Biodiversity* (Wilson 1988) was the spark igniting wider interest and usage. Biodiversity as theme, title and keyword has burgeoned ever since. The idea, perhaps its intuitively obvious meaning, expanded its impact beyond the scientific community. The contents of *Biodiversity* are revealing. The themes of the 57 chapters extend beyond pure science. Six main threads reoccur. Eleven chapters provide inventories and estimates of the extent and types of life on Earth. Ten explore loss rates, historic and recent. Ten examine the value, financial, aesthetic and for planetary health, both as current uses plus future potential. Eleven analyse economics, financial structures, forces and approaches especially as agents of destruction. Twenty-eight chapters discuss the policy, practice or examples of conservation. Eleven chapters range much more eclectically through social systems, religion, poetry and environmental politics, reflecting attitudes to biodiversity. An epilogue pulling together the main messages concludes that 'maintaining global biodiversity seems to depend on the collective behaviours and perceptions of people toward their habitat'. E.O. Wilson, the editor, ends his introductory chapter with the suspicion that the success of conservation will be decided by an ethical response, that 'the fauna and flora of a country will be thought part of the natural heritage as important as its art, its language and that blend of achievement and farce that has always defined our species'. (See plate 1.)

Plate 1 *The O'Hara farm, Galway. The O'Hara family abandoned their farm during the Irish Potato famine. Mr O'Hara ended up a match-seller on the streets of New York and never saw his children again. The failure of potato crops was due to dependence on one genetic variety vulnerable to disease. The lack of genetic biodiversity cost the O'Hara family dear*

The wide scientific, social and philosophical horizons incorporated by the concept of biodiversity are evident in much of the early literature (see Table1.1.). The linkage of economic and political issues with biological science was important. Conservationists could point to the real financial value of biodiversity and conservation widened its horizons to address the underlying human pressures threatening life on Earth.

Table 1.1 *Major themes within biodiversity. The presence of six main themes in texts from the conceptual origin of biodiversity in 1980 prior to the 1992 UN Convention on Biodiversity*

Paper or publication	Lovejoy 1980	Norse and McManus 1980	Norse et al. 1986	Biodiversity 1988	National Science Board 1989	Biological Diversity and developing countries 1991
Theme inventory	Definition. Global species total.	Definition. Global species total.	Definition. Global species total.	Definition. Global species total. Habitat specific totals.	Definition. Global species total. Gaps in knowledge.	Definition. Global species total.
Loss rates	Estimated losses due to forest destruction.	Causes (settlement, transport, fragmentation, agriculture, forestry, over-exploitation, introduced species).		Global and habitat specific estimates. Geological patterns. Case studies.	Causes (human population growth, habitat loss, ecosystem function loss).	Habitat examples. Causes (development, market failure, interventions, habitat loss, overexploitation).
Value	Potential uses.	Potential uses (food, energy, chemicals, raw materials, medicine).	Products (actual and potential). Ecosystem services. Forestry. Psychological well-being	Medicine, industrial, food, potential.	Uses of micro-organisms and plants.	Use and non-use values, potential, habitat examples.
Economics	Valuation of ecosystem function.		Timber production.	Market failure to price biodiversity, new economic approaches, pricing biodiversity.	Economics of utilisation (clothes, food, medicine, shelter).	Economic approaches to value biodiversity.

Table 1.1 continued

Conservation	Gene banks, botanical gardens, species specific schemes, ecosystem management, legislation.	Concepts. Conservation and management. of forests and rare species. Laws.	Priorities, case studies, technologies, reintroduc- tions.	Recommenda- tions for research on inventories, education, training and funding in developing countries.	Aid policy procedures for priority sites, habitat examples, criteria
Attitudes	Philosophical, psychological, affinity to life, right to exist.	Ethics and stewardship.	Green movement, society links to ecosystem, morals, religion, philosophy.		Ethical value of biodiversity.

The IUCN had promoted the idea of a global convention on biodiversity since 1981 and in 1987 the United Nations Environment Programme (UNEP) called for an international treaty to govern conservation and sustainable use of biodiversity. Preliminary meetings led to the creation of the Intergovernmental Negotiating Committee for the Convention on Biological Diversity and by May 1992 a final version of the Rio Convention had been drafted. The years of preliminary meetings reflected political tensions surrounding biodiversity. Developing countries, many harbouring the greatest concentrations of biodiversity, were keen to retain full control and rights to exploit these resources. Some developed countries promoted an awkward mixture of biodiversity as everyone's global heritage, over which the impecunious developing world was expected to stand guard for all our sakes, coupled with a desire that industries should have free rein to explore and develop new products from the treasure trove. Developing countries were fearful that any discoveries would be whisked away to foreign laboratories and reappear, maybe in synthetic form, ringed about by strict patent controls.

The Rio Convention consists of 42 articles covering definitions, principles, research, education, finance and administration. Article 1 summarises its objectives:

> The objectives of this Convention, to be pursued in accordance with its relevant provisions, are the conservation of biological diversity, the sustainable use of its components and the fair and equitable sharing of the benefits arising out of the utilisation of genetic resources, including the appropriate access to genetic resources and by appropriate transfer of relevant technologies, taking into account all rights over those resources and to technologies, and by appropriate funding.

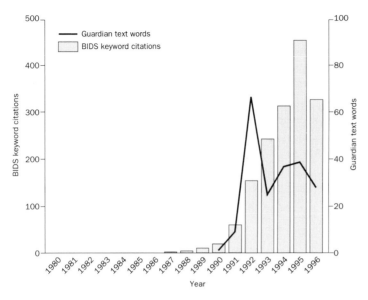

Figure 1.1 *The appearance and increasing usage of the term biodiversity. Frequency in the scientific literature is shown by numbers of articles containing biodiversity in their title or abstracts, derived from the Science Citation Index, BIDS (1996 data up to September). Use in the mainstream media is recorded by frequency of appearance in articles in the* Guardian *newspaper (1996 data up to June)*

The world discovers biodiversity

Definitions of biodiversity crystallised in the scientific and technical literature throughout the beginnings of the 1980s. Early discussion on biodiversity embraced social, economic and cultural horizons. This was something more than 'boffins' offering up insights to a baffled and disinterested public. The concept of biodiversity was attuned with a popular mood. The imagery of nature's richness, imminent threat and human dependence on the continued welfare of the natural world was embraced rapidly and globally by the public, while providing a formal framework which integrated emerging research themes for environmentalists.

The speed with which awareness of the term spread and the intuitive response of a wide audience reflects a coincidence of environmental concerns and campaigns during the 1980s. These were the same concerns that prompted the original scientific synthesis of biodiversity and provided a receptive, informed public. Four mutually reinforcing trends came to a head by the late 1980s:

- environmental campaigns involving publicity, direct action and political activism, particularly linked to tropical rainforests;
- public awareness of, interest in and familiarity with the richness of wildlife via the media;
- environmental threats such as acid rain, ozone depletion and global warming that highlighted the interdependence of natural and human systems, the role of healthy ecosystems for the maintenance of planet Earth and the global reach of threats;
- interest in biological resources whether prompted by biotechnology in the developed world or sustainable exploitation in developing countries.

Environmental campaigns

Environmental campaigning organisations have blossomed since the mid-century. The World Wide Fund for Nature (WWF, started as the World Wildlife Fund) is typical of post-war international conservation bodies, combining research, funding conservation projects, political lobbying and public awareness. From the start WWF had a clear publicity agenda, fronted by a famous television presenter, Peter Scott, and intentionally built around a core membership of influential princes and millionaires. The WWF was launched with a six-page, shock-horror newspaper article in the UK's *Daily Mirror* linked to the plight of East African game, already seen as a tourist resource. Latterly the WWF has been rattled by charges of complacency and internal schisms as an autonomous US WWF grew apart from the European core. None the less the WWF's Giant Panda motif is recognised internationally.

The 1970s and 1980s saw the rise of headline grabbing activism, most famously Greenpeace and Friends of the Earth, both started from small protests in 1969. Greenpeace graduated from a small demonstration closing a US–Canadian border road in 1969, via a first sailing into a US nuclear test zone in 1971 to a string of publicity grabbing actions. From the start Greenpeace had an overt media agenda. Swashbuckling physical bravery, often pitted against the media unfriendly monoliths of government, big business and televisual baddies such as Soviet and Japanese whalers earned huge public admiration (see Plate 2). The litany of success is remarkable. Incursions into French nuclear test zones in 1972–3 resulted in arrest. Smuggled film of French brutality to the prisoners discredited official denials of mistreatment and, by association, claims for the safety of tests. In 1975 the first anti-whaling mission resulted in film of harpoons fired over the heads of Greenpeace inflatables into whales. In 1977 anti-seal cull action in Newfoundland combined survival in atrocious Arctic weather with a visit by film star turned animal rights activist Brigid Bardot ('blood, death and sex', ideal). In the 1980s they took to steeplejacking, using climbers to scale chimneys and factories to unfurl condemnatory banners. In 1983 a daring landing in Russia ended in a ten-hour boat chase towards Alaska. The Greenpeace boatman was captured but hidden film was later retrieved. Ex-Greenpeace staff developed equally adventurous variations, for example, *Sea Shepherd*, a reinforced trawler used to hunt and ram (and perhaps bomb) pirate whalers; the Environmental Investigation Agency, using undercover agents, often wired for film and sound to expose illegal trade in animal products such as fur and ivory; and Lynx, a trade organisation most famous for its anti-fur poster depicting a woman dragging a coat, entitled 'It takes 50 dumb animals to make a coat but just one to wear it'.

In the developing world campaigners fought a combination of government and business, at best corrupt and indifferent, at worst murderous. Tropical rainforests became a potent symbol (see Plate 3). From Brazil and South East Asia the fight against wanton clearance, often allied to rights of indigenous peoples, attracted global publicity. The rainforest campaigns are credited with bringing native people to the centre of conservation. In Brazil encroachment of settlements and forest clearance, often illegal and enforced by private armies, had been highlighted since 1980 by the

Plate 2 *Greenpeace anti-whaling protest. Physical bravery and brilliant imagery caught the public imagination, reinforcing a growing concern for the diversity of life on earth*

Photograph Greenpeace/Stone

National Rubber Tappers Council which was formed to defend the rights of all those making a sustainable living from forest resources. Headed by the charismatic and internationally renown Chico Mendes, the Council's campaign culminated in a 1987 stand against eviction from an estate recently purchased by ranchers. Despite death threats and killings the Tappers held their ground. The Brazilian government was moved to create an extractive reserve allowing sustainable use of the forest. In December 1988 Mendes was shot dead by men he had credited as organisers of local death squads. By 1992 and the looming Rio Summit their trial was still dragging on with an increasing sense that the system would save them. Meanwhile Brazil was shocked by the international outcry and the first, albeit shaky, national environmental agency was set up.

In 1989 the first extensive publicity began for native Amerindians who were fighting loss of territory to loggers and huge dam projects. (See Plate 4.) International visits by exotically clad chiefs and mass rallies in Brazil made for good television. From 1986 tropical forest campaigns had been bolstered by international umbrella organisations (even if split between anti-trade, direct action versus dialogue, co-operation approaches). Direct forest losses were compounded by massive dam projects which often used billions of dollars of development aid to build huge dams which destroyed habitat and displaced people, all too often bringing few benefits. Throughout the late

Plate 3 *Intact tropical rainforest. The photogenic quality of rain forests was an instrumental factor in the rise of environmentalism in the 1980s. This forest is on Mount Kupe, Cameroon (see Chapter 4)*

1980s campaigns halted projects around the world, culminating in protests at a 1989 World Bank meeting in Berlin credited with shocking complacent bankers to review their funding of projects. The rainforest campaigns brought together so many biodiversity themes: genetic, species and ecosystem loss; the importance of ecosystems for local need and global health; the role of indigenous people; under-valuation of natural resources; the potential uses of medicinal and food plants reflected in local knowledge.

Media impacts

Rainforests are also deeply photogenic. Public interest in wildlife, especially the exotic, has an ancient history from amphitheatre spectacle, royal gifts, menageries and zoos. The appetite for natural history television appears insatiable, reinforced by high quality magazines such as *National Geographic* in the US, *Animals*, relaunched as *BBC Wildlife* in the UK, and *Geo* in Australia. Wildlife media developed rapidly following World War II. The famous BBC Natural History Unit was founded in 1957 on the back of increasing success and provided a non-sentimental, expert coverage of the natural world. By the late 1980s the developed world had a culture of wildlife television with personalities such as Jacques Cousteau and David

Plate 4 *Human impact on rainforest. Logged forest but regenerating, South America*

Attenborough, famous to the point of caricature. Attenborough's 'Life on Earth' series became a touchstone for the richness of life on Earth. Most famously his romp on the forest floor with a family of wild gorillas left a strong sense of kinship and threat. A massive, fascinated and knowledgeable audience has been exposed to the literal diversity of life as never before. A night seldom goes by without TV film of animals eating each other, mating, or sometimes both at once. The media impact can also be a source of problems. Superb documentaries have been accused of creating an image of a world still overflowing with life and a taste for the exotic so that more familiar local animals and plants are ignored. The media have also been a major factor in creating a sense of global anxiety. We are all instantly aware of threats to planetary health.

Global threats

The third coincident of the late 1980s was that the public, so supportive of green activists and familiar with wildlife, witnessed the discovery of threats to planetary health that brought a deeper understanding than the familiar, grim but often rather local problems from oil spills. Acid rain, ozone depletion and global warming combine similar themes. Each are complex and diverse processes but at their core is pollution of the atmosphere by human activity: acidic nitrogen and sulphur emissions from burning of fossil fuels; Chlorofluorocarbons (industrial coolants) destroying high level ozone, and greenhouse gases, particularly carbon dioxide and methane. In each case the chemical degradation provokes cascading complications. Acidic precipitation affects soils, forest health and freshwater life with economic losses to wood and fisheries. Human health is also at risk from increased dissolved metals and nitrogen compounds in water. Ozone depletion resulted in increased ultraviolet radiation reaching the earth's surface, potentially damaging to most life and known to suppress marine phytoplankton production with possible losses to fisheries. Human health is at risk from increased skin cancers. Global warming (still much debated but increasingly accepted) promises a host of impacts. Global shifts in climatic zones bring associated shifts in natural and agricultural productivity, spread of tropical disease vectors, extremes of weather and sea level rises. Acidification, ozone loss and warming all suggest that damage to natural systems is not only a symptom but provokes further problems. All three are international in scope, encouraging a sense of global community and responsibility. For all their complexity the catchy titles were readily accepted. These problems combined ecological science, economic threat and environmental politics, just as biodiversity.

Biodiversity's potential

The final coincidence was an increasing awareness of the potential of biological resources, many yet to be discovered. The value of tropical forest species (mainly plants but also animals, fungi and microbes) for medical and food uses became an important argument used to defend forests. Some examples were simply known forest products that had always been used locally but could now be marketed effectively, pandering to adventurous tastes in the developed world. Suggestions included the Lulo (*Solanum quitoense*), a fruit from Columbia and Ecuador especially good as a juice and Amaranth (*Amaranthus* sp.), a protein rich crop once used by the Aztecs but suppressed by Spanish colonists due to religious symbolism. This is now marketed as a health food in the USA. Improvement of existing crops through discovery of new species or strains has been promoted through several famous examples. A fourth species of Maize (*Zea* spp.) was discovered by chance in Mexico in the late 1970s. Maize is one of the world's five most important crops. In 1972 botanists in Peru chanced upon a weedy new tomato. Seeds were collected but overlooked for several years. Genes from this species boosted soluble solids (the tasty component) of tomatoes by up to 50 per cent, potentially benefiting the US tomato industry by millions of dollars. Wild strains of crops may enhance resistance to disease, pests and climatic variation. (See Plate 5.)

Why biodiversity matters

The conceptual origins of biodiversity, rooted in reviews of the state of the global environment, show that the subject has always been more than an academic puzzle. Ecology, economics and environmentalism each brought an awareness of the

Plate 5 *Knock-on effect of rainforest logging: logged rainforest replanted with cannabis, South America. Pressures from the developed world are often cited as a cause of environmental destruction in developing countries. Here part of the damage is exported back*

significance of biodiversity. Expert and popular opinion contributed to a sense that biodiversity mattered and therefore so did our wise stewardship of the planet. Even opposing views, for example, the clash between developing countries keen to utilise biodiversity and the developed world insistent upon its conservation, merely high-lighted this. The importance of biodiversity was explicit in Norse and McManus' (1980) paper with a review section entitled 'Material benefits of biological diversity' covering food, energy, medicine, raw materials, genetic discoveries but also ideas of human empathy with wildlife and philosophical attitudes'.

The greatest insight, something humans had always sensed but only made explicit since the 1980s, is the importance of **ecosystem function** for planetary health. Ecosystems are made up of their living inhabitants and physical components. The activities of species driven processes, called **ecosystem services**, such as water and gas cycling and purification, formation of soils, growth of food, fuels and products. Not only are these services vital to humans, they are free. An ecosystem function is the sum total of its services. Ecosystem functions depend on the organisation of the ecosystem, the diversity of taxa, their abilities and activity and distribution. Individual services are particularly dependent on species that have unique abilities (e.g. fixing nitrogen) or play a key role in processes without necessarily any special characteristic (e.g. keystone predators). Taxa with similar roles and abilities have been dubbed **functional groups**, acting as **drivers** of the ecosystem processes. These are so new and poorly understood that it is easy to draw the wrong conclusion, that if only we knew which the important driver functional groups and species were we could afford to lose the rest. The real lesson is exactly the opposite. We do not know and there is emerging evidence that ecosystem functions and their ability to recover from degradation improve with increasing diversity of species. All biodiversity is important.

In addition the importance of biodiversity touches all areas of our lives. Table 1.2 summarises specific historic, current and potential examples.

Table 1.2   *Historic, current and foreseeable importance of biodiversity to humans*

	Historic	*Current*	*Foreseeable*
Ecosystem services	Pollination of crop plants by bees.	Wetlands to clean pollution. Effective as buffers against many pollutants, especially nutrient over-enrichment.	Soft coastal engineering against sea level rises from global warming.
Medicine	Penicillin, antibiotic derived from fungus.	Horse Shoe Crab blood used in bioassays of toxins.	Rainforest plants. New drugs, to combat existing and new diseases.
Biotechnology	Rattans, Asian forest palms used for fibres for many building and artifacts.	Bacterial genes introduced to crops to confer resistence to insects.	Metal digesting bacteria. Potential use to clean contamination and extract valuable metals.
Environmental monitoring	Losses of bird species as evidence of DDT impact.	European forests damaged by acid precipitation.	Species range and habitat changes in response to global warming.
Food	Irish potato famine due to over-reliance on one vulnerable cultivar.	The weedy tomato, wild gene improved domestic cultivars.	Several high value exotic rainforest fruits.
Recreation	Victorian Fern craze, widespread hobby.	Birdwatching. Major recreational activity in developed world.	Eco-tourism. Growth area for global tourism.
Pets and domestic animals	Exotic zoo animals as gifts and status symbols.	Value of current pet trade.	Potential from use of historic breeds as we rediscover useful attributes.
Political and social	Symbolic function in Celtic religion; animals as companions to St Cuthbert.	Eco activists, e.g. Greenpeace and Earth First; highly visible political impact.	Local ownership and control of resources in developing world.

Not only does biodiversity exist as a resource we currently use or can foresee a use for. Biodiversity is also an insurance. We are increasingly aware of how much we do not know. Our stewardship of biodiversity is vitally important not just to us but for the future. The concept of biodiversity may have a short history, but the history of life on Earth teaches a telling lesson in uncertainty, of serendipity.

The next section explores the history of life and lessons from the past.

Biodiversity through time

The importance of time

Our awareness of current threats to biodiversity and prospects for future survival arises in part because we know of animals and plants that have become extinct in the past. News media feed an appetite for tales of dinosaurs, mammoths, living fossils and missing links. There is an appreciation that the wildlife with which humans share the planet today is not the same species as in the past. The biodiversity of Earth is not a static phenomenon. Processes that work over huge time scales (hundreds and thousands of millions of years, called **deep time**) have spawned species and ecosystems today that are conspicuously different to those of the past, e.g. different species, dominant taxa, new habitats colonised. Yet life today is intimately linked to the past by evolutionary lineages. Everything has an ancestor. Today's species may be different but many lifestyles they lead are recognisably similar to those of extinct creatures. Many habitats are similar. Some were changed and some were dead but overall the diversity of life around us is simply the continuation of the biodiversity of the past.

To understand life today we need to know its history. To conserve biodiversity into the future we need to check for any lessons from the past. We need to know if extinctions today are qualitatively different to natural losses throughout the history of life, comparing the types of diversity at risk today versus those wiped out in the past. We need to know if current losses are quantitatively different, if more species are being lost, at a faster or slower rate, over a larger or smaller area. The history of biodiversity may provide insights to the causes of extinction, the vulnerability of different types of species, what are the danger signs, which survived – all useful for the strategy and tactics of conservation. There may even be warnings or reassurance for what might happen if some extinctions are now inevitable. The future of biodiversity depends on unravelling the past.

The history of life

How long has life existed on Earth?

Planet Earth is about 4.550 billion years old. A sense for this achingly long time grew only slowly throughout the nineteenth century as geological and biological enquiry unravelled the pattern and processes of the planet's natural environment. The 4.550 billion figure derives from the decay of radioactive forms of elements in rocks. The decay rates of **isotopes** (slightly different forms of individual elements) are precise. The time required to produce the ratios of different isotopes found in rocks today can be calculated, establishing how long these rocks have existed. Moon and meteorite rocks, created at the same time as planet Earth both predict the 4.550 billion age of the solar system. (See Figure 1.2.)

Era	Period titles		Million years ago (x 10^6)
CENOZOIC	Quaternary	Holocene	
		Pleistocene	2
	Tertiary (T)	Pliocene	
		Miocene	
		Oligocene	
		Eocene	
		Paleocene	
			66
MESOZOIC	Cretaceous	K	144
	Jurassic	J	208
	Triassic	Tr	245
PALAEOZOIC	Permian	P	286
	Carboniferous	C	360
	Devonian	D	408
	Silurian	S	438
	Ordovician	O	505
	Cambrian	C	570
PRECAMBRIAN (pC)	Proterozoic		2500
	Archean		4500

Figure 1.2 *The geological calendar. Note the time spans of periods are not drawn to scale*

In Western culture the scientific revision of Earth's history replaced an understanding of the world through a literal interpretation of the Book of Genesis. This revision left a serious question. The known history of life on Earth, evidenced by visible fossils, extended as far back as the start of a geological period, the Cambrian, 570 million years ago. The fossil record bursts into life, the quantity and variety of animals so marked that this appearance is dubbed the Cambrian explosion. Earth showed no signs of life for the preceding 90 per cent of its history. The eruption of life was so abrupt that Charles Darwin, writing *Origin of Species*, regarded the lack of any signs of life beforehand as a serious challenge to his theory of evolution. Discoveries since the 1960s have cast a startling new light on the history of biodiversity, sounding ominous warnings but also strange reassurance.

The oldest rocks left to us are 3.8 billion from Greenland with other remnants between 3.6 and 3.36 billion in South Africa and Australia. In 1977 a palaeontologist took samples from a South African remnant, 3.4 billion years old. The hard, flint-like rock was called a chertz, formed by mineral laden volcanic water poured over thick mud. Microscopic examination of very thin sections of the chertz showed minute round, occasionally dumbbell shaped structures. From their size, shape and apparent division (the dumbbells) they were eerily like modern day bacteria to look at. This discovery prompted searches of other ancient rocks, most famously Fig Tree formations (South Africa, 3.36 billion), Gunflint (Canada, 2.5 billion) and Bitter Springs (Australia, 1.1 billion). All contained microfossils of single-celled life, with increasing diversity in younger rocks. The Gunflint samples yielded spheres, rods, filaments, stalked blobs and tentacled forms. The very oldest rocks known do not harbour microfossils. However, the ratio of carbon isotopes C12 and C13 in the rocks is unusual, with C12 higher than expected if **abiotic** (non-living) inorganic chemistry processes were solely responsible. Elevated C12 may be another sign of life due to preferential use of this isotope by photosynthetic organisms.

The overriding message of these rocks is that life goes back through deep time as far as we can delve. From perhaps 3.8 billion years ago Earth has supported life. This early origin and continued survival in one form or another suggest that Earth is very conducive to life. The planet is near enough to an external energy source, the Sun, to prevent freezing, but not too close that elements evaporate and the surface is destroyed. Earth is not so small that the atmosphere is stripped away, not so large that the atmosphere is dense and muffling. The conditions allow carbon-based chemistry to flourish. Carbon is abundant and reactive, without being dangerous and the basis for the most complex and varied chemistry. Condensed water is plentiful. Water is the best solute for most chemistry and liquid across a range of temperatures, fast enough to allow chemical reactions but not so fast that systems collapse. In addition water's surface tension and heat exchange properties have been exploited by life.

The early origin of life and continued survival despite terrible natural disasters is profoundly reassuring. Life is robust and resilient. Although humans have wreaked havoc with some habitats nothing we have done has threatened the overall survival of life. Other lessons are more challenging. For 3.1 billion years, 80 per cent of the history of life, biodiversity was dominated by bacteria, **cyanobacteria** and other life forms dismissed today as microbes, as if size alone was the criterion for judging the importance of biodiversity. Judged by their monopoly of deep time the biodiversity of Earth is the story of the bacteria.

Biodiversity rising

The history of biodiversity is an epic adventure. Throughout there are patterns and crises that have a particular relevance for our understanding of the patterns and crises today.

The age of bacteria

From their ancient origins to 670 million years ago biodiversity was dominated by bacteria and their kin, a **kingdom** of life called the **prokaryotes**. Prokaryote evolution is the foundation for the diversity of life today, genetic, metabolic and ecological. Their role in Earth's history shows the power of living things to alter the environment at a planetary scale.

The precise environmental chemistry of the **Archean** is disputed but general characteristics were an atmosphere of nitrogen, hydrogen, carbon dioxide, water vapour and traces of other gases with the exception of oxygen. The lack of oxygen affected ancient seas, creating reduced conditions which determine the forms of chemical ions present. Earth's atmosphere today is predominantly nitrogen with about 20 per cent oxygen. This archaic environment provided the cradle in which prokaryotes developed a diversity of metabolic systems for three purposes: releasing energy, building carbon reserves, and tolerating extreme environments. This variety of metabolisms and tolerances still exists and at a fundamental level represents a much greater diversity than the difference between a jellyfish, an ant and a human, all of which share the same energy processing metabolic system.

The Archean spawned microbes able to survive in a range of frighteningly severe habitats, viciously alkaline soda lakes, thermal vents discharging water hotter than 100°C and desiccating salt pans. Many of these especially tolerant bacteria belong to a group called the **Archeabacteria**, separated from the other bacteria, called the **Eubacteria**, the two **domains** combining to form the prokaryotes. All other life belongs to a third domain, the **Eucarya**, often called the eukaryotes. These three domains are distinguished by the biochemistry of structures inside their cells called ribosomes, essentially sites where genetic code is read and used to build protein. This three-domain divide, Archeabacteria, Eubacteria and Eukarya is the fundamental division of biodiversity on Earth.

The diversity of bacterial metabolism reflects the lack of gaseous oxygen in the Archean atmosphere. Free oxygen is toxic to many extant bacteria, their metabolic chemistry disrupted by this highly reactive chemical. Oxygen is highly reactive, readily combining with common elements. Atmospheric oxygen is a massive departure from chemical equilibrium, its very presence evidence for life, perturbing geochemical cycles. Free oxygen's origins are also from life, an example of biodiversity changing global environments and the first and most severe biodiversity crisis to afflict Earth.

Oxygen is the waste product from photosynthesis splitting water to obtain hydrogen ions. Cyanobacteria are credited with this atmospheric revolution. Fossil evidence for Cyanobacteria goes back to 2.6 billion. From then until about 1.8 billion oxygen production was mopped up, reacting with abundant ions such as calcium and iron in the sea. Thick Iron Band deposits some 2.0 billion years old are evidence of particularly rich deposition. As iron and other chemical sinks for the oxygen were used up atmospheric oxygen increased. The replacement of a carbon dioxide–methane rich greenhouse atmosphere with oxygen may have sparked the first known Ice Age. For many prokaryotes oxygen was a deadly poison. Global ecosystems were destroyed. Remnant communities were banished to habitats beyond oxygen's reach such as waterlogged mud and the deep sea. For others this was an opportunity. Respiration using oxygen to release energy (aerobic) is more efficient than anaerobic. Prokaryotes able to use oxygen thrived. In the wake of the oxygen holocaust the third domain, Eukarya, appeared, the result of an extraordinary biodiversity link up.

Eukaryotes differ from prokaryotes by possession of discrete, membrane bound structures inside their cells called **organelles**. Different organelles specialise in different tasks. Two of the most distinctive are **mitochondria**, the site of aerobic respiration, and **chloroplasts**, the site for conversion of sunlight radiation to stored chemical energy. Both chloroplasts and mitochondria have their own genetic information, separate to that of the rest of the cell. When the cell divides the mitochondria and chloroplasts duplicate themselves simultaneously but separately. These two organelles are very like stripped down bacteria, retaining their specialist metabolic skills but harboured in the bigger cell. Mitochondria and chloroplasts appear to be remnants of once independent bacteria, subsumed into a third bacterium, the main body cell. In addition the motile hair-like cilia outside many animal cells are uniquely similar to the structure of motile spirobacteria. There are still hints of this origin of Eukaryotic cells, termed endosymbiosis, among modern creatures. One of the most extraordinary arrangements is *Mixotricha paradoxa*, superficially a single-celled animal found in the hind guts of termites but actually a collection of at least five types of pro and eukaryote. (Box 1). In biodiversity terms eukaryote cells, yours and mine, are an intimate synergy of different bacteria, individual identities sunk into a co-operative venture, combining the diversity of metabolic systems. Every eukaryote is a bacterial homunculus.

So the fundamental biodiversity of life is three domains, Archeabacteria, Eubacteria and Eukarya. Prokaryote diversity is characterised by their varied metabolic skills. The origin of atmospheric oxygen was the result of prokaryote photosynthesis; even the smallest most ancient life was capable of modifying the planetary environment. A global biodiversity catastrophe due to oxygen toxicity and climate change occurred but even this most terrible event did not wipe out life. Oxygen tolerant prokaryotes took over and new cellular organisms developed from an intricate association of prokaryotes.

Lilos and hallucinations – nature experiments with animals

Darwin fretted at the lack of Precambrian fossils. In 1947 on an Australian mountain range called **Ediacara**, formed of rocks some 700,000,000 years old, a Precambrian fauna was discovered.

The Ediacaran fauna is now known from similarly aged rocks around the world, a global diversity of shallow marine habitat animals but one that does little to explain the Cambrian explosion. The Ediacaran ecosystem is represented by beautifully preserved fossils of entirely soft bodied animals. If animals they be, most resemble frisbees, ferns and rings. There are no hard structures, no mouths, no sense organs, no limbs. The flat discs or lobes are segmented, much like an inflatable airbed. The Ediacaran animals look like a natural experiment into life as a lilo (Box 2). Initial interpretations placed the fossils as precursors of known animal lineages such as worms or anemones. An alternative is that the Ediacarans are an entirely separate lineage, an experiment with no obvious features to justify links to later groups, a novel ecology. Their flat and fronded shapes may have relied on the large surface area relative to volume to obtain nourishment and exchange gases, rather than internal

Box 1

Mixotricha paradoxa, *a creature or community?*

Mixotricha paradoxa (the Latin translates as the mixed hair paradox) lives in the hind guts of an Australian termite, *Mastotermes darwinensis* (see Figure 1.3). Many termites rely on an exotic gut fauna to digest wood, the flora including many strange prokaryotes and single-celled eukaryotes. *Mixotricha* appeared to be just another gut ciliate but detailed analysis of the structure reveals a multiple symbiosis. The pear-shaped body is a eukaryote cell, a Trichomonad protistan. Inside the cell are bacteria which may function in place of mitochondria, which the creature lacks. On the surface are three types of spirochaete bacteria, the most numerous (500,000) which undulate to move the cell, a second bacterium which is part of the anchor mechanism for the small spirochaetes, and a few larger spirochaetes. Other example of spirochaetes forming motility associations has been found among other termite gut flora.

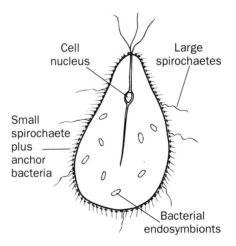

Figure 1.3 Mixotricha paradoxa

Source: Redrawn from Margulis 1993.

organs. Quite what they are remains a source of debate. One recent suggestion is that many may be lichens. They are animals utterly alien in design and ecology.

As suddenly as they appear and dominated so the Ediacaran fauna was lost. The Precambrian/Cambrian boundary is uncertain. From Ediacara to the Cambrian explosion lies a 100-million-year span of so-called 'small shelly fauna': odd scales, spines and tubes, perhaps the skeletal structures once attached to softer bodies or perhaps complete cases and bodies of another strange experiment. Ediacara, the small shelly fauna and the Cambrian explosion may represent three quite separate developments, nature experimenting with how to be an animal, each dominant within the 100-million span, but only the Cambrian fauna leaving any descendants. The biodiversity of animal life today is the progeny of this third fauna, the Cambrian experiment.

Box 2

The Garden of Ediacara

The strange plant-like shapes of many Ediacaran animals gave rise to an image of the community as the Garden of Ediacara, an echo also of the Garden of Eden at the dawn of obvious life.

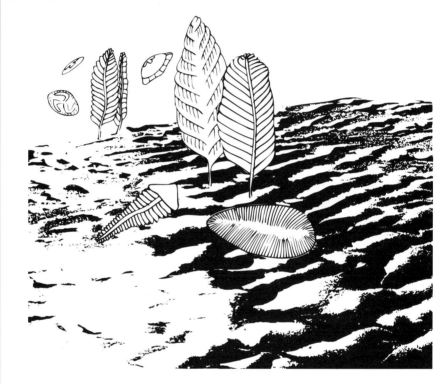

Figure 1.4 *Ediacaran animals*

The sudden diversity of animal life that had perplexed Darwin appeared to contain representatives of all the major animal lineages alive today. Ever since the Cambrian animal diversity has been variations on the same themes. Although animal diversity today may seem extraordinary, the majority of species are developments of but one design, the insect, and mocked as '600 million million years of fussing'. The image of an ever-increasing diversity of animal life with its latent messages of progression and improvement towards the pinnacle that is humanity has also been undermined by analyses of the early Cambrian faunas, suggesting that some fossils represent animal designs that have left no descendants. One fundamental level used to describe animal biodiversity today is that of different **phyla**, e.g. chordates, arthropods, molluscs, the different phyla defined by fundamentally unique body designs. Some Cambrian fauna may represent entirely separate phyla now completely lost. So, the history of animal diversity since the Cambrian may be one of fundamental loss of phyla but increasing species diversity within a handful of those remaining.

This sense of loss is epitomised by the Burgess Shale animals, an early Cambrian marine fauna exquisitely preserved in shale deposits in the Canadian Rockies. The Burgess Shale fauna appear to be whole communities apparently swept to their doom from shallow continental shelves by mud slides. Many Burgess Shale animals can be allotted to known groups. Others do not fit any recognised animal phylum. The Burgess Shale fauna may be a snapshot of yet another natural experiment with animal designs. This revision of early biodiversity casts a darker shadow on our understanding of life. There are no obvious hints as to the likely survivors or the doomed, winners and losers. Several of the most heavily armed creatures fit no known groups while one of the least offensive, *Pikaia*, may be an early chordate. Survival of lineages may be a matter of luck, as much as any deterministic, predictable pattern. Today's biodiversity is contingent on this early lottery with no guarantee that, if the natural experiment could be run again, the same outcome, the same survivors, the same biodiversity would arise. No animal seems a better example of this extraordinary fauna and status of weird wonders in or out of known phyla than *Hallucigenia*, described in Box 3.

These early faunas and their fates send important messages. Initial animal biodiversity suggests experimentation with biodiversities now lost. Different fauna waxed and waned. The fate of individual lineages may be unpredictable. Today's animal diversity is but massive variation on a few themes. To mangle Ford's famous maxim on car colour, you can have any animal you like so long as it is an insect. Animal diversity is no more than one outcome from a mass of permutations. No lineage has a guarantee. Nature is not sentimental to her firstborn.

Founding dynasties and terminal disasters

From the Cambrian to the end of the Permian (the Palaeozoic, 345 million years long) global biodiversity increased in two major ways: first, the diversity of animal species, primarily in the seas; second, the colonisation of land opening up new ecosystems.

Throughout the Palaeozoic marine faunal richness increased, measured as families or genera, even allowing for artefacts such as greater availability of younger rocks. Different animals dominated at different times, Trilobites in the Cambrian, Nautiloide molluscs in the Ordovician, fishes in the Devonian. Important ecosystems became established such as coral reefs. The reef-building taxa were different to those today but the type of ecosystem is similar. Ways of life became established, e.g. filter feeding by clams and brachiopods in the Ordovician. Arms races were won and lost as predator and prey locked into escalating spirals to overcome one another.

The late Ordovician witnessed the first land plants and invertebrate animals: the centipede/millipede/insect line, spiders, woodlice. Some of the earliest land plant communities, including the famous *Rhynia*, show damage that resembles attacks from herbivorous invertebrates. The establishment of land plants was based on an association with fungi, sharing nutrients, a co-operative mutualism that is still the foundation of terrestrial vegetation diversity. Land opened up new opportunities. The forest created habitat, physical space and new resources. The predominance of insect species diversity today depends on groups intimately co-evolved with plants. While these modern insects and the plants appear later the terrestrial insect–plant link was

Box 3

Hallucigenia sparsa, *weird wonder or upside down worm?*

In 1977 The journal *Palaeontology* contained a paper describing an animal named *Hallucigenia* in honour of its baffling form. Pictured tiptoeing over the Cambrian ooze on stiletto spines, the description ending coyly 'its affinities remain uncertain'. The article speculated on the lifestyle this 2–3 cm long monstrosity, the concentration of individuals around remains of a worm only adding to the eerie quality, 'the spines could have been embedded in the decaying flesh'. In 1985 a review of the Burgess Shale had *Hallucigenia* listed under miscellaneous animals, by now perhaps 'grasping food with its tentacles'. With the publication of Stephen Gould's *Wonderful Life* (1989) *Hallucigenia* became an icon of all that seemed mysterious about the Cambrian fauna, of the 'stunning disparity and uniqueness of anatomy'. Yet Gould was also wary that so strange a beast may be reinterpreted. By 1991 published evidence of other wormlike fossils from the early Cambrian, with obvious legs but also bearing spines and plates, forced a revision of *Hallucigenia*. It is now generally regarded as allied to the phylum Onychophora, represented today by velvet worms, mysterious inhabitants of the tropical rainforest floor. Gould could not help a pang of disappointment, writing, 'I regret the loss of a wonderful weirdo'.

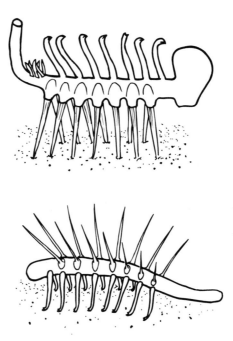

Figure 1.5 Hallucigenia sparsa; *stilt walking alien from Conway Morris, 1977; reinterpreted as spikey worm from Ramskold and Xianguang, 1991*

established in the Palaeozoic. The end of the Palaeozoic is marked by a global catastrophe, the Permian extinction that resulted in the massive diversity losses, over 95 per cent of all species. This crisis has been tied to continental drift driving the southern land masses into one supercontinent, Gondwanaland. This resulted in

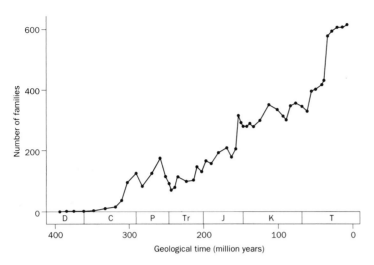

Figure 1.6 *Increasing insect family diversity through time. The drop at the end of the Permian probably represents real decline, the rapid rises in the Tertiary (T) are artefacts of good preservation*

Source: Redrawn from Heywood 1995.

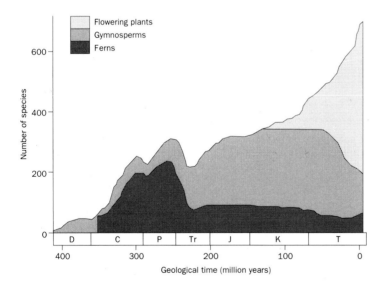

Figure 1.7 *Vascular plant diversity since the Devonian period. Overall diversity increases though different groups dominate at different times*

Source: Redrawn from Groombridge 1992.

massive losses of shallow marine habitats as the continents collided and climatic change due to the single mass of land. Recovery from the Permian extinctions was the start of a contest between different groups of vertebrates to dominate terrestrial habitats, resolved (for the time being) 179 million years later at the end of the Mesozoic with the extinction of the dinosaurs. (See Figure 1.6.)

During the early Triassic five dynasties of large animals vied for the land: labyrhinthodont amphibians, three derived from reptilian stock (Rhyncosaurs, Therapsids, dinosaurs) and Synapsids ('mammal like reptiles'). Each dominated at different times, but the sequence defies the prejudice inherent in classifications that hint at mammalian superiority. The ultimate winners were the dinosaurs, dominating the large terrestrial animal niches from the late Triassic to end of the Cretaceous. Terrestrial biodiversity toward the end of the Mesozoic would have looked largely familiar to us. There existed broad-leaved and coniferous woodlands with butterflies and birds flitting from flower to flower and small mammals, frogs and toads in the undergrowth. Only the occasional dinosaur and absence of grass would be strikingly different. (See Figure 1.7.).

At the end of the Cretaceous the dinosaurs and many even more ancient lineages were completely wiped out. In 1980 a massive meteorite impact was proposed as the cause

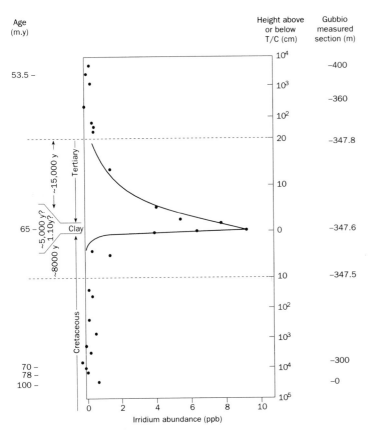

Figure 1.8 *Simplified diagram of the 'Iridium anomaly spike' from Alvarez et al. 1980 paper proposing a meteor impact as cause of the K/T boundary extinctions. Iridium abundance is shown from rocks near Gubbio, Italy, before, at and after the K/T boundary along with rock depths and times*

Source: Redrawn from Alvarez *et al.* 1980.

by father and son team Luis and Walter Alvarez and colleagues Frank Asaro and Helen Michel. This idea was prompted by discovery of high concentrations of the element iridium in deposits at the Cretaceous–Tertiary (K/T) boundary outside the town of Gubbio in Italy. (See Figure 1.8.) Iridium is very rare on the Earth's surface but abundant in meteors. The elevated levels in the Italian samples suggested debris deposition following a meteor impact. The drama of this scenario provoked fierce debate, partly responsible for our increasing awareness of the extent and nature of extinctions throughout the history of life and comparison with current threats to biodiversity. The sudden catastrophic hypothesis is now widely accepted but then the cause remains uncertain, perhaps a meteor strike, perhaps massive volcanic activity.

In 1996 the journal *Science* reported that a fragment of the actual asteroid may have been found. The 3 mm chip of rock was at the very base of sediments apparently laid down by the K/T impact, its position out in the mid-Pacific also hinted at an impact hitting at a shallow angle, all the more destructive, rather than burying itself in the Earth. In addition to the global iridium other geophysical clues support a massive impact, see Box 4. The impact hypothesis has prompted a serious search for other likely asteroid or comet threats to Earth. The Spacewatch scheme in the USA estimates numbers of very dangerous 1 km+ size at between 1,000 to 2,000. Analysis should have identified any serious collision courses by 2008. In the meanwhile in 1994 an asteroid passed by Earth at only 105,000 km.

The contest between vertebrate dynasties to dominate the large animal niches on land again shows the turnover of biodiversity. Extinctions and radiations, both natural processes, are equally important. The history of land vertebrates is one of

Box 4

Evidence for the K/T impact hypothesis

There is now multiple evidence for the impact hypothesis.

Iridium Anomalously high levels of iridium have now been found at over 100 K/T boundary sites worldwide.

Shocked quartz Quartz grains with distinctive shock-induced stress patterns, even as far as molecular rearrangements. Only found at known meteorite impact sites and requiring extreme conditions to create. Subsequently found at K/T boundary sites.

Mass carbon deposits Unusually large deposits of charcoal and soot at K/T boundary sites suggestive of massive global fires ignited by the impact.

Spherules Spherical particles formed by cooling of molten droplets. Droplets found in K/T boundary layers do not fit chemical make-up and form of similar droplets that volcanoes may produce.

Diamonds Diamonds from K/T boundary with carbon isotope ratios similar to those from meteorites not terrestrial sources.

Amino acids Similar analysis of amino acids from K/T boundary material showed similarity to those known from meteorites, not terrestrial sources.

Tidal wave debris Thick jumbled deposits of material, typical of massive tidal wave, particularly around the Caribbean.

Chicxlub crater In 1990 the 200–300 km wide, 65-million-year-old Chicxlub crater at the north tip of the Yucatan peninsular was widely publicised as the impact site. In addition this massive crater lines up in an arc with several smaller craters scattered over the globe and dated to the same time, suggesting a multiple or glancing impact. (See Figures 1.9 and 1.10.)

shifting dominance. The Mesozoic world was generally warmer than today, a greenhouse world benefiting the dinosaurs, versus the icehouse currently favouring mammals. If global climate changes markedly the ecological balance of advantage may shift, though dinosaurs as such are long gone. The dinosaurs' fate has prompted some of the hardest revisions to our idea of extinction, once so firmly tied to ideas of inferiority and obsolescence. The danger may be one less of bad genes and more of bad luck.

How many species have there been?

In spite of all the extinctions and losses life has never been utterly destroyed. The diversity of life with which we share the planet can be reduced to a mundane equation: species diversity today = speciation – extinctions. What we have left is a surplus and, although species diversity has generally risen over time, the majority of species that have ever lived are now extinct. Estimates put the current animal and

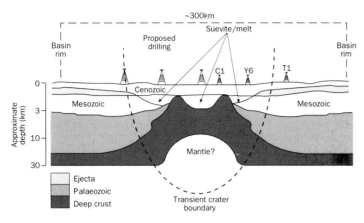

Figure 1.9 *Simplified cross-section of Chicxlub crater site, Mexico. The unusual rock structures were revealed by geological drill sites (C1, Y6, T1) used for oil exploration. The mantle, Palaeozoic and Mesozoic rocks are severely disrupted. Suevite and melt rocks are various products of catastrophic disturbance. Cenozoic rocks have subsequently reburied the site*

Source: Redrawn from Smit J. (1994) Blind tests and muddy waters, *Nature*, 368, 809-810.

plant species diversity at 2–5 per cent of the historic total. Such calculations rely on historic evidence for species diversity and longevity, data most reliably gleaned for marine invertebrates that have the most complete fossil record. Individual species vary greatly in lifespan, many fossil mammals lasting only 1 million years, the average for marine invertebrates 11 million years, for all species taken together 5–10 million years. Combine estimates for the known existence of animal life, average duration of species, the numbers of animals alive today and adjustment for declining background extinction rates and one estimate of the numbers of animal species that have ever lived comes out at 18,375,000.

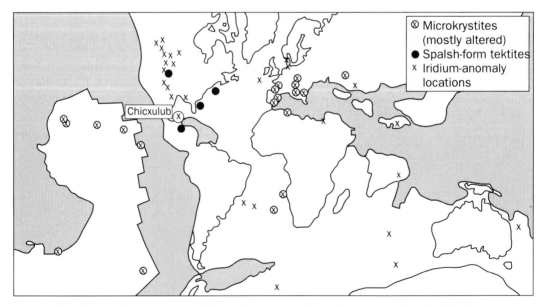

Figure 1.10 *The continents at the end of the Cretaceous when a giant meteor hit. The Chicxlub crater site is marked along with sites from which fall-out debris such as high Iridium contamination, splash-form tektites (droplets of glass) and microkrystites (crystallised droplets of rock) have been found*

Source: Redrawn from Smit J. (1991) Where did it happen? *Nature*, 349, 461-462.

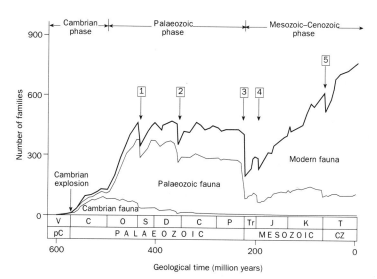

Figure 1.11 *Animal family diversity since the Cambrian explosion. Diversity generally increases but different fauna dominate different periods and there are occasional mass extinctions. The five mass extinctions are marked: 1 Ordivician; 2 Devonian; 3 end Permian; 4 Triassic; 5 end Cretaceous (K/T)*

Source: Redrawn from Heywood 1995.

Natural extinctions

Types of extinction

The natural fate of all species is extinction. This can mean the total loss as a species dies out. Alternatively as a species evolves individuals will eventually become so different to the ancestral species that it is effectively no more.

The lifespans of genera have been tracked through the fossil record. Most genera have comparatively short lifespans, a few longer, a very few very long. Most genera share three characteristics: few species, few individuals in each species and small geographic ranges that together can account for short lifespans of species. All species encounter good and bad years. A very bad run may wipe out a species. In any set of species for every few doing very well there will be some doing very badly, to the point of oblivion. This gradual attrition wipes out species by nothing more than a simple statistical probability. Raup distinguishes three modes of extinction. The **field of bullets** describes random extinction unrelated to relative fitness of species, the name stems from an analogy of soldiers advancing across a battlefield being mown down at random. **Fair game** extinctions are selective losses of less fit species, the classical Darwinian survival of the better adapted. **Wanton extinction** is selective of species but not related to relative fitness or adaptation to environment.

A continual attrition is perfectly natural, some species lost to nothing more than the statistics of bad luck. Such losses have been labelled **background extinctions**. They are eclipsed by the sudden, global, **mass extinctions**, pivotal in the history of life. The nature of mass versus background extinctions is revealing.

Background versus mass extinctions

From the mid-1970s a group of American scientists, notably John Sepkoski and David Raup, compiled a detailed compendium summarising known origin and extinction dates for families of marine animals. The work coincided with the revived interest in mass extinctions and catastrophes following the K/T meteor impact hypothesis. The database was a treasure trove for measures of variations in extinction rates, in particular comparison of background versus mass extinction rates. The work revealed some striking patterns, notably peaks of extinction every 26 million years with five especially serious extinction events between the Cambrian and K/T boundary. The periodicity inspired speculation of dramatic causes, mostly extraterrestrial, whether meteorites, super novae or an unknown death star dubbed Nemesis.

Mass extinction rates appear catastrophic. There have been five major mass extinctions. Analyses provide stark estimates of species losses. In percentage terms estimates are Late Ordovician 84–85 per cent, late Devonian 79–83 per cent, end Permian 95 per cent, end Triassic 79–80 per cent and the K/T event 70–76 per cent. Nonetheless most extinctions have happened outside of these catastrophes, the result of background extinctions. The fossil record suggests an average species lifespan of about five million years, for any one species equivalent to a 20 per cent chance of extinction every one million years. Expressed as the probability of a species going extinct per year this is 0.0000002. Assume current estimates of species diversity on Earth today are correct at between 10 and 30 million, this background extinction rate predicts yearly losses of 2–6 species. Background extinction rates show signs of a steady decline over time for some taxa, notably marine invertebrates, but perhaps increases for others, e.g. land plants. A declining rate may reflect the resilience of surviving taxa, winnowed by previous losses. Extinction rates vary with taxa. Differences of lifespan between taxa may result from their biological traits. These attributes may be the now familiar factors of range and abundance but other characteristics can be important. Analysis of extinction rates for marine Gastropod molluscs (snails and their kin) shows that the feeding ecology of the larvae is a better predictor of susceptibility to extinction than adult feeding. Taxa characteristics of ecosystems with a history of vulnerability such as coral reefs may always be at risk. (See Figure 1.12.)

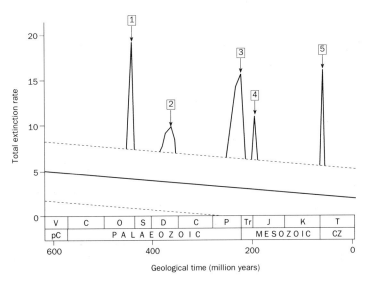

Figure 1.12 *Extinction rates of marine animal families since the Cambrian period. Extinction rates are given as numbers of families lost per million years. The solid line shows the average rate for background extinctions, which all lie within the dotted line boundaries. The five major mass extinctions are revealed by spikes of increased extinction rates*

Source: Redrawn from Heywood, 1995.

To separate background from mass extinctions may be an artificial divide of a seamless gradient of varying rates. Mass extinctions affect many more species, impact globally and are, geologically, short events, but these are simply increases in rate and scale rather than qualitatively different from background losses. Mass extinctions may be nothing more than the extreme end of a continuum, the very rare end of the skewed distribution. Mass extinctions are better defined by their impact than quantities such as area afflicted, numbers lost, time span (Box 5).

Box 5

Characteristics of mass extinctions

Qualitative losses Major lineages entirely wiped out. Several wiped out simultaneously. Many habitats affected, marine and terrestrial. Disparate taxa affected. Ecological stage reset via ecological collapse as dominant taxa are replaced.

Scale Global impacts, taxonomically broad, rapid.

Survivorship Luck? Survival of taxa may be pure good fortune or due to a characteristic that turns out to be fortuitously useful but could not be a prescient adaptation to impending danger. Such adaptations, beneficial in retrospect, are called exaptations.

Luck continued Taxa that survived previous mass extinctions survive subsequent events.

Range Ability to adjust geographic range rapidly or be widespread already.

Tolerance The ecology of taxa may make them largely immune to a particular event. Freshwater and deep-sea fish survived the Devonian mass extinction that hit hard the shallow marine taxa.

Mass extinction events coincide with major changes to the global environment but no one type of event can be matched to all mass extinctions. Three of the five extinctions between the Cambrian and K/T boundary have been linked to meteor impact evidence, but the biggest extinction, the Permian, does not. The Late Devonian mass extinction comprised losses to tropical marine ecosystems, including shallow reefs and open water. Deeper and cooler water taxa survived. The evidence suggests global cooling as a cause. The Permian extinction saw massive losses to shallow marine and terrestrial ecosystems. There is some evidence for a series of extinctions in rapid succession. By the end of the Permian all land masses joined into one supercontinent, Pangea, reducing shallow marine habitats, allied to deeper oceans as mid-oceanic sea bed forming ridges were less active and deepening sea basins pulled water off the continental shelves. Global climate changed to arid terrestrial systems. The K/T boundary extinction is now credited to a massive meteor impact causing regional havoc and global change.

Biodisparity, a qualitative and quantitative measure of the extent of extinctions

Given the complexity of extinction patterns, simply measuring numbers of species lost cannot properly depict the impact on biodiversity. Loss of Burgess Shale weird wonders or dinosaurs represents a qualitative void. **Biodisparity** describes this sense of the variety and extent of loss, be it of lineages, lifestyles, anatomy or genes. Biodisparity is a tricky new area, relying on quantifying the ecology and morphology of taxa than rather just totting up total numbers. Biodisparity can describe the kaleidoscope of biodiversity in a region or time. The history of life on Earth shows that winners and losers are unpredictable. Conservation should depend as much on monitoring the spectrum of biodiversity rather than the fate of a handful of apparently precious taxa since we cannot second guess what life forms will inherit the future.

Extinctions, rebounds and radiations

Even the Permian mass extinction did not completely wipe out life. In the wake of such catastrophes ecosystems have recovered and rebounded, often repopulated by a burst of evolutionary activity characterised by the diversity of taxa and speed of change, called **radiations**. New species take up ecological roles previously occupied by the recently extinct. Rebounds and radiations vary in rate and outcome regionally and there can be a lag period after the extinction suggesting ecological processes require some convalescence. The occasional major losses of biodisparity mean that radiations develop from an increasingly limited stock of fundamental lineages.

The history of life is as much about extinction and loss as diversification and creation but extinction and environmental change today are widely seen as symptoms of a problem, not a natural process. What evidence is there for a biodiversity crisis on late twentieth century Earth?

The current crisis

Miners' canaries

In 1973 a new species of frog was discovered living among the rock pools of montane forest streams in the Conondale and Blackall ranges of Queensland, Australia. While trying to move a captive frog Queensland Museum staff were astonished to see the beast vomit up six wriggling tadpoles. Incredulity greeted reports that this new species, soon named the gastric brooding frog, *Rheobatrachus silus*, nurtured its tadpoles in the female's stomach. The tadpoles survive in the potentially lethal stomach environment due to massive changes to stomach structure and function, switching off the digestive processes. In 1980 $23,000 was invested into pharmacological research to look at the inhibitory mechanisms with a view to human medical uses. Soon more was known about this frog's morphology than any other Australian species. Meantime the frog became a potent symbol for campaigns to protect the Conondale ranges which are increasingly threatened by logging, habitat fragmentation and aquatic pollution. Local and international recognition of the danger to this remarkable amphibian saw its use on T-shirts and car stickers. A three-year study into the impact of logging was commissioned to begin in 1982, building on intensive surveys between 1976 and 1981. This frog was a classic example which combined so many themes of biodiversity: a previously unknown species of the tropical forest; direct interest for biomedical benefit; subject of local and international conservation campaigns. This was a frog with a future. (See Figure 1.13.)

Figure 1.13 *Car bumper sticker using the Gastric Brooding Frog as a flagship species for conservation, in this case against the threat from logging of forests*

Source: Redrawn from Taylor M.J. (ed) (1983) *The Gastric Brooding Frog.* Croom-Helm, London.

The last wild gastric brooding frogs were seen in 1979. In 1980 and 1981 none were found. The gastric brooding frog is now regarded as extinct.

The loss of this frog may be a natural background event, unusual only in so far as we witnessed its passing. However, in the last twenty-five years extinctions of amphibians have drawn attention to such losses. Examples have occurred globally and from different habitats. The golden toad (*Bufo periglenes*) of Costa Rica, famed for the brilliant orange colour of the males and intensively studied from the early 1970s crashed to apparent extinction between 1988 and 1990. Common species also show population declines, e.g. western spotted frog (*Rana pretiosa*) now extinct from a third of its range in the USA. These losses of both the intensively studied and common came to worldwide attention following an American National Research Council Workshop on declining amphibian populations, held in Irvine, California, in 1990. The workshop explored four themes.

1 Evidence of amphibian extinctions, population and range declines from around the planet.

2 Evidence that such losses represented a genuine event that could be distinguished from natural (background) population fluctuations.

3 Amphibians as indicators of general environmental degradation.

4 Identification of any single cause.

Despite problems in distinguishing unusual loss rates from background events, compounded by media coverage including suggestions of frog-napping by space aliens with a taste for amphibians, evidence for the extent of the declines was convincing, including species apparently protected in pristine sites. However, there was little evidence for one single cause. Habitat loss, fragmentation, introduced predators, disease, pollution, acid rain, pesticides and consumption by humans have all been linked to losses. Local damage may be reinforced by synergistic processes with factors such as UV radiation and global warming. Amphibians may be an example of what have been termed **miners' canaries**, from the use of caged canaries down mines which would keel over in the presence of poisonous gases, giving the miners time to escape. The amphibian losses are one of several miners' canaries, an advanced warning of impending danger. Amphibians might be very good candidate canaries. A permeable exposed skin and open eggs readily interact with the environment. Complex lifestyles risk disruption at many stages from different factors. Reliant on freshwater at least occasionally their survival reflects wider catchment health. The consensus was that the recent declines were a remarkable event, not over and above the existing biodiversity crisis but a very clear signal of environmental degradation.

The accolade of miner's canary was widely popularised by Niles Eldredge's (1991) book. His exploration of the reality of recent extinctions opens with the observation that migratory songbird numbers have declined in the northern hemisphere. Eldredge suggests that these birds fulfil the same role on a global scale as the canaries of old. Songbird losses in North America echo the message of the frogs, except that one ultimate cause may be apparent. US Breeding Bird Census returns show local extinctions, population decline and range contractions of warblers, thrushes and flycatchers. These birds are forest specialists. The damage is done by fragmentation of woodland into patches below a critical size threshold. Small woodland remnants were exposed to predators and the egg parasite Cowbird, which hunts along the woodland edges, scouring the whole patches more effectively.

Extinctions of colourful frogs and birds attract the attention of conservationists and public alike but these were not the only startling losses in recent decades. Evidence from northern Europe points up similar problems afflicting fungi, particularly those that appear as mushrooms. These fungi live symbiotically with tree roots. Fungal filaments, mycorrhizae, sheath the roots, releasing scarce soil nutrients in return for carbon products. The fungi are vital for woodland health. From Germany, the Netherlands, Austria and Czechoslovakia the results of regular fungal forays from the early 1960s report losses of between 40–80 per cent of species, with severest declines in the 1980s. There have been economic consequences for humans. The chanterelle (*Cantharellus cibarius*) harvest in Saarbrucken, Germany, fell from 6,000 kg per year in the 1950s to less than 200 kg a year in the 1970s. Extinctions, population declines and range contraction are exactly the same symptoms as for frogs and birds. Air pollution causing increased nitrogen inputs to the soil has been blamed. Trees may dispense with their expensive symbionts once nutrients increase or the fungi may be more sensitive to change.

Whole ecosystems may also respond as canaries, most famously the bleaching of coral reefs and collapse of temperate fisheries. Corals are members of the animal phylum Coelenterata, along with jellyfish and anemones. Coral reefs are built by corals that can secrete a calcium carbonate exoskeleton. Not all corals build massive reefs. Those that do are restricted to the tropics with average monthly sea temperatures no lower than 20°C all year round. The variety of corals is matched by the richness of other life dependent on the reef. Yet all this massive ecological engineering depends on a symbiosis of coral polyps with single-celled algae, called Zooxanthellae, living inside the cells of the polyps' guts. Photosynthesis by the algae provides most of the corals' carbon supply and polyp waste is recycled. In return algae receive nitrogen and phosphorus nutrients scarce in the surface waters of the tropics from food snared by the coral. The photosynthesis also creates an alkaline local environment allowing the calcium carbonate accretion (recent evidence suggests that perhaps the algae may be the dominant partner controlling calcification to enhance photosynthesis). Corals without symbiont algae can build skeletons but at terrible energetic costs and massive reefs are impossible.

Coral bleaching results from the loss of zooxanthellae. Algal cells can degenerate or be expelled. In extreme cases polyp tissues containing algae detach completely. The corals lose their colour and die leaving the bleached, white skeleton. Knock-on effects include loss of fish species due to habitat destruction and blooms of toxic algae, with economic losses rippling through to humans. Bleaching has been known since the 1930s as a response to local distress but from 1979 massive regional bleaching episodes have been recorded in every year (except 1981 and 1985) with peaks every three years. Many causes have been suggested: disease (no evidence); oxygen toxicity (but oxygen production tends to decline at bleaching episodes); UV light (but some bleach beyond UV penetration depth); pollution (but pristine sites have bleached); and heat plus intense light (some evidence for local impacts). One factor is consistently associated with mass bleaching, raised sea water temperature. A mere 1°C increase over the normal long-term warm season average is sufficient to induce bleaching. Bleaching has been cited as evidence of global warming. Projections of sea temperature rises of 1.5 to 5°C over the next one hundred years have been suggested. Rising sea levels and increased storms would be an additional burden to reefs. The bleaching episodes since 1979 may be an early warning, a whole ecosystem miner's canary, the first to register the effect of global warming. (See Figure 1.14.)

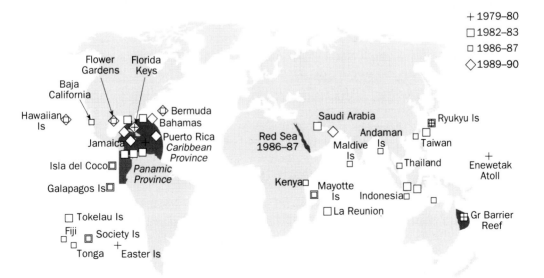

Figure 1.14 *Global mass coral bleaching incidents from peak years, 1979–90*

Source: Redrawn from Glynn (1991).

Losses of taxa and ecosystems in crisis are serious problems in their own right but the coincidence of events over the last two decades, their acceleration and the underlying message of planetary health in collapse as impacts cascade through natural systems setting off further disruptions suggests a deeper crisis. The miners' canaries have become too ominous and spectacular to ignore. The damage to natural ecosystems is no longer seen as merely symptomatic, evidence of damage already done, but as feeding back to precipitate yet more danger.

An end to evolution

We can all name an extinct animal. An awareness of imminent extinction is a familiar theme dogging the tail-end of most wildlife documentaries. The history of life on Earth is as much a tale of extinction as any other process. Extinction is so familiar, so natural that it begs the question, what are we worried about? Our concern may stem from a faddish environmentalism or an understandable but self-centred desire for planet Earth to remain much as it is, despite the haunting message of change from the ghosts of lost worlds. Extinction is not the problem.

The threat to biodiversity arises if extinctions due to human activity are qualitatively and quantitatively different to natural events. In addition the opportunity for natural rebound and recovery may no longer exist. Recent extinctions take place so rapidly, may be so widespread and cause such disruption to ecosystems that a global catastrophe snowballs. This combination of losses with the ferocity of a mass extinction with no chance of recovery has been termed the end of evolution (Ward 1995). Citing the Permian mass extinction as the first major event and the K/T catastrophe as the second, Ward believes a third event, starting with glaciation some 2,000,000 years ago and accelerated by humanity is already underway. The combination of a climatic first strike (glaciation) causing broad disruption followed up by global damage wrought by humans to surviving ecosystems as analogous to a massive meteor impact is frightening.

Other scientists disagree and believe losses at the end of the twentieth century are not much different to natural rates. The idea has achieved mythical status. One critic has dismissed tales of a current mass extinction, as a case of scientists not letting the evidence get in the way of such an important idea. The very word biodiversity can be abused. In 1996 Professor Obel, Kenyan Chief Scientist to the President, marketed a cure for HIV/AIDS, called Pearl Omega, at £350 a bottle under a company name of Biodiversity. Launched with backing from the state press, Obel commented 'we have hit the jackpot!', while dismissing calls for evidence that it worked, its nature and despite a ban on sales from the Health Ministry. Pearl Omega was withdrawn from sale later in the year following international press coverage. In the developed world biodiversity has become such a popular buzz word that articles increasingly sneer at this jargon.

The seriousness of recent extinctions can only be judged against past patterns. We need to understand what ecosystems do and their importance, if any, for general planetary health. We need better estimates of how much life shares the planet and the world's special biodiversity hot spots, accurate measures of current losses other than gaudy telegenic fauna and how much we can afford to lose before the crisis becomes terminal. We need to unravel the patterns and processes of biodiversity.

Summary

- The term biodiversity was coined in the 1980s and rapidly popularised, combining aspects of science, economics and politics.

- Life on earth is an ancient phenomenon. Earth is highly conducive to life. Life has survived massive global crises. Most species that have lived are now extinct but life has always recovered from previous crises.

- Recent species extinctions and ecosystem collapse suggest a new crisis, equal in scale to historic mass extinctions, may be taking place.

Discussion questions

1 Why did the concept of biodiversity catch the public imagination in the 1980s?

2 What messages, worrying or reassuring, does the history of life on earth carry for biodiversity today?

3 Are there any miners' canaries in your country?

See also

Definitions of biodiversity, Chapter 2.
Current extinction rates, Chapter 4.
International conservation treaties, Chapter 5.

General further reading

Wonderful Life. Stephen Jay Gould. 1989. Hutchinson, London.
Gripping account of the Burgess Shale fauna and wider lessons for our interpretation of the history of life on Earth.

Extinction. Bad Genes or Bad Luck? David M. Raup. 1993. OUP, Oxford.
Review of historic extinctions, including mass extinctions, their causes and frequency.

The Miner's Canary. Niles Eldredge. 1992. Virgin Books, London.
Historic and recent extinctions and evidence for current crisis.

The End of Evolution. Peter Ward. 1995. Weidenfeld and Nicolson, London.
Extinctions, historic and recent, and biodiversity.

'Changes in biological diversity'. T. E. Lovejoy. 1980. In G. O. Barnay (ed.) *The Global 2000 Report to the President, Vol. 2 (the technical report)*, pages 327–332. Penguin Books, Harmondsworth. Founding paper describing biological diversity. Includes estimates of extinction rates as forest habitat declines.

'Ecology and living resources biological diversity'. E. A. Norse and R. E. McManus. 1980. In *Environmental Quality 1980: the eleventh annual report of the Council on Environmental Quality*, pages 31–80. Council on Environmental Quality, Washington D.C. Founding paper describing biological diversity. Wide horizons explicitly include philosophical and social topics.

Biodiversity. E. O. Wilson (ed.). 1988. National Academic Press, Washington D.C. The book of the 1986 National Forum on Biodiversity.

'Section 4. Biodiversity and ecosystem functioning: basic principles. Section 5. Biodiversity and ecosystem functioning: ecosystem analyses'. 1995. H. A. Mooney, J. Lubchenko, R. Dirzo and O. E. Salsa (eds). In V. H. Heywood (ed.). *Global Biodiversity Assessment*. CUP for UNEP, Cambridge. Very thorough, detailed and effective review of this major topic.

2 The creation of biodiversity

Patterns of biodiversity are created by ecological and evolutionary processes. Lessons from ecology and evolution are important for conservation. This chapter covers:

- **Origins and remits of ecology and evolutionary science**
- **Ecological patterns and processes**
- **Evolution and the diversification of species**

Biodiversity is the outcome of evolutionary and ecological processes. True but glib. This explanation does not describe how these processes work, disguises serious disagreements over the relative importance of different factors and, since aspects of evolution and ecology are part of biodiversity, is circular: biodiversity begats biodiversity. The university libraries where I work stock 1.5 km of shelves stuffed with books and journals dedicated to these topics. The purpose of this chapter is to provide a secure foundation from which you can explore these themes without feeling as if you are swimming through blancmange. This chapter dovetails evolutionary and ecological pattern and process where they are relevant to understanding biodiversity and explains what they are. First, the scope of evolutionary and ecological sciences and their respective territories needs definition before you can venture into those kilometres of library shelves.

The **ecological** domains are defined by the **population, community** or **ecosystem** and their interactions with the environment. The **phenotype** is central. Relevant ecological processes range from the autecology of individuals through to whole ecosystem function. The patterns and processes have an impact on genetic diversity but as a consequence. Ecological patterns and processes within communities and ecosystems may ignore the genetic domain which is manifested by species. The evolutionary domain is defined by a focus on genetic processes, patterns and consequences and the variation they create. The **genotype** is central. Relevant evolutionary processes range from adaptation (as genetic selection in response to the environment) through microevolutionary changes within a species to macroevolution, speciation and extinction. These changes may be driven by the environment but the genetic diversity remains the theme, linked to the species. Some genetic processes, e.g. mechanism for mutation and variation, could function without any environmental influence.

Evolution, ecology and the origins of confusion

The origins of ecology and evolutionary science

A fascination with the origins, patterns and causes of nature's variety can be traced back at least as far as Greek texts, most famously Aristotle's (4th century BC) *Historia Animalium* listing over 500 species, followed by Theophrastus (4th to 3rd century BC) with an *Historia Plantarium* describing plants and their uses. From the start attitudes to nature split between a vision of purposeful, harmonious systems versus nature as nothing more than individualistic bits linked like a machine. These classical writers may be an artificial horizon, a sort of Cambrian explosion of surviving texts. Ancient cave paintings (e.g. Vallon-Pont-d'Arc) depict unintelligible but haunting murals of humans and animals and hint at an awareness of the interactions of nature in humanity's archaic history. Medieval and pre-Enlightenment scholars maintained a robust interest in nature's variety, interactions and importance. The *Summa Theologica* of St Thomas Aquinas, in the thirteenth century, analyses the orderliness of nature, including the sentiment that could be drawn from the most recent biodiversity texts on ecosystem function: 'It is better to have a multiplicity of species than a multiplicity of one species.' Other insights were less useful, biodiversity included many dubious mythical wonders. (See Plate 6.)

Plate 6 *Biodiversity promotes healthy ecosystems. Using sophisticated indoor growth chambers, called Ecotrons, ecologists at the Centre for Population Biology have shown that a diverse mix of plant species recycles gases and water more efficiently than a similar biomass of few species*

Photograph COI/CPB.

The European Enlightenment revolutionised attempts to classify the natural world. New classifications used the physical resemblance of animals and plants to catalogue, to create an order, in keeping with the growing intellectual confidence of the age. Older magical, symbolic, sometimes medical categorisations were lost. Many Enlightenment systems were firmly grounded on a vision of divine creation with species as immutable types, any new species at best filling up preordained pigeon holes. Most famous is the work of Carl Linnaeus, a Swedish biologist, who devised the system for applying, classifying and naming species that is still in use today.

The themes of other natural historians are familiar in biodiversity research today. In the 1700s John Harris published *Lexicon Technicum*, describing natural history as 'a description of any of the natural products of the earth, water or air, such as beasts, birds, fishes, metals, minerals, fossils, together with such phenomena as at any time appear in the material world, such as meteorites'. In 1749, Georges, Comte de Buffon, Keeper of the Royal Zoological Garden in Paris, started publication of his *Histoire Naturelle*, intended as an inventory of the Earth and its wildlife, highlighting patterns of geographical distribution, adaptation and migration, a global biodiversity assessment of its day. The role of species in maintaining a healthy environment was documented: for example, the Reverend Gilbert White writing on earthworms in letters later published as *The Natural History of Selbourne*, 1789; Richard Bradley an eighteenth-century horticulturist commenting on the value of birds for pest control. The interdependence of species was understood, for example, John Ray, writing *The Wisdom of God Manifested in the Works of the Creator*, 1691.

The accumulation of knowledge continued through the early nineteenth century. Finally came the publication in 1858 of Charles Darwin's and Alfred Wallace's theories of evolution by natural selection, informed by the distributions of living species, fossil record and a mechanism for evolution driven by day-to-day contest for limited resources. The only major component missing was an understanding of genetics. Even if not yet christened evolutionary and ecological science appeared to have had a very sound gestation, but the reverse was true.

Modern ecology: illegitimate origins and confused childhood

The disparate heritage of ecology caused problems in the definition of the legitimate territory of this science. The term ecology (then spelt oecology) was coined by the German zoologist Ernst Haeckel in 1866, in part to categorise the topic as a mere aspect of physiology and to distinguish it from biology. Ecology was an awkward combination of the traditions of natural history (everything from grotesque menageries through to the ideas of Darwin) and the other major thread of nineteenth-century natural biology, physiology. At various times both traditions subsumed or abhorred the newcomer. In 1893 Burdon-Sanderson, President of the British Association for the Advancement of Science, distinguished ecology as one of the three divisions of biology (physiology and morphology were the other two), with ecology defined as 'a philosophy of nature'.

Ecology is not a radiation of ideas from a secure central core, nor a synthesis of ideas into a greater whole. It is an umbrella of concepts. Three blights (no secure pedigree, multiple parents and a subject matter unloved by the great biological themes of the late nineteenth century) got the subject off to a shaky start.

A fourth source of confusion, in part a reaction against this sense of illegitimacy, soon developed. During the early twentieth century there was an emphasis on making field studies rigorously scientific to bolster ecology's status and to understand nature all the better to master and improve upon it. There was a desire to make ecology coherent by developing a quantitative, mechanistic, reductionist approach, as if nature could be stripped down like a big machine, echoing the rigours of physics and chemistry. Ecology had transferred its interest from explanations dominated by history (the legacy of biogeography, dispersal and evolution) to explanations that forgot time. Ecology as a science had no clear past and now ignored historical time as a process. G. E. Hutchinson's famous paper of 1959, 'Homage to Santa Rosalia or why are there so many kinds of animals', is a revealing example of the shift in emphasis. Specifically addressing global animal species richness, Hutchinson lists factors to consider for a theory to predict the total: food chains, plant/animal interactions, complexity and stability of ecosystems, productivity, size, environmental rigour, competition and the environmental mosaic. The paper omits historical ecology.

Ecology was riven by rival schools of thought – **population/community ecology** concentrating on species and populations versus **ecosystem ecology** analysing the cycles and flow of energy and resources with a sense of regulation and balance. Alternative visions included communities as integrated **super-organisms** versus **individualistic assemblages**. This contest between ecological systems as benignly self-regulating or driven by selfish contest echoed the Enlightenment and even the Greek debates between a vision of nature driven by purpose versus trial and error. An important source of confusion is the different scales across which ecology and evolution can be studied. Different processes are important at different scales. Disagreements between different schools have been fearsome, though the acrimony is often out of the public gaze at conferences. A published flavour of the debate can be found in the eminent journal *American Naturalist* in 1966. In 1960 Hairston *et al.* published an article analysing the structure of communities, in particular the relative importance of factors affecting wildlife with different modes of feeding (decomposers, herbivores, carnivores). In 1966 Murdoch published a critique that rubbished their ideas. In a brief footnote immediately after Murdoch's paper, Hairston and Smith wrote: 'In spite of an extensive exchange of views with him we remain in complete disagreement.'

Then, just when ecology was trying to look grown up by talking in numbers, increased public awareness of environmental problems in the 1960s saw the term ecology purloined as a general social and political creed. Many scientists were wary of or annoyed by the blurring of their science with environmental politics. The emergence of a science of biodiversity has to some extent healed this rift.

The ecology of biodiversity

The search for patterns and processes

Robert MacArthur studied under Hutchinson, driven by the same question as his mentor: why are there so many species of animals? MacArthur's work became a dominant influence in late twentieth century ecology, immortalised in the MacArthur–Wilson Theory of Island Biogeography (the same Wilson that edited 1988's *BioDiversity*). In 1972 MacArthur's book *Geographical Ecology* was published, written when terminally ill. The introduction opens: 'To do science is to search for repeated patterns'.

Later, Chapter 7, entitled 'Patterns of species diversity', begins with the frightening litany of patterns, tropical to temperate, land to sea, plant and animal and frets: 'Will the explanation of these facts degenerate into a tedious set of case histories or is there some common pattern running through them all? A very brief review of explanations that naturalists have proposed will help put the discussion in perspective.'

This section faces the very same challenge to produce a coherent review of the many ecological factors creating diversity and controlling its distribution: first, a hierarchy of factors differentiating three broad types of ecological factors, primary influences of time and space; then templates such as physical habitat structure or patterns common to food webs; and third intimate processes depending on individual species' actions and attributes.

Primary factors, geographical and physical

Primary ecological factors are large-scale global or regional physical influences. They determine the general environment within which the other ecological factors create local, intimate detail or disruption.

- History and age. Biodiversity is generally greatest in oldest ecosystems.
- Gradients. Biodiversity changes across environmental gradients (e.g. latitude, altitude, depth, aridity and salinity).
- Area. Biodiversity increases with increasing area.
- Isolation or islandness. Biodiversity decreases with increasing isolation.

Ecological templates

Biodiversity at regional or smaller scale suggests patterns or structures to which species diversity adheres or is driven to fill up. Structures may be literal, e.g. the physical architecture of a habitat, or apparent, e.g. rules governing how species link together in food webs.

- Productivity, available energy. Globally biodiversity increases with productivity but locally increased or decreased productivity typically lowers biodiversity.
- Scaleless patterns. Ecological factors such as area or productivity have a measurable scale (e.g. diversity per hectare or relative to solar energy). Scaleless patterns include the consistent ones found in food webs and patterns of species richness relative to population abundance and body size.
- Habitat heterogeneity and patchiness. The greater the variety of types of habitat, the greater the diversity of species.

- Habitat architecture. Diversity increase with expanding architectural complexity of the physical habitat.

Interactive intimate factors

Primary factors and ecological templates typically promote or reduce biodiversity in a predictable way. Biodiversity increases with increasing area, or decreases with decreasing habitat complexity. The third set of factors can enhance or suppress diversity, depending on the spatial and time scales. They are characterised by the local intimate scale at which they operate, directly dependent on the attributes, behaviours, interactions and susceptibilities of the species involved.

- Succession. Successional processes comprise the arrival, establishment and replacement of species through time. Changes may be driven by environmental agents and species altering habitats.
- Interactions between species. Interactions between species influence short-term, local diversity and power evolutionary diversification as winners, losers, exploited and exploiters struggle for dominance.
- Disturbance. Local physical destruction, as against regional, climatic variations linked to gradients, may destroy diversity. Equally, disturbance may be frequent enough to prevent a limited gang of species monopolising a habitat but not so frequent that everything is wiped out.
- Dispersal and colonisation. Dispersal (active or passive) and colonisation (arrival plus survival) depend on the attributes of individual species and represent the species interface with the primary factor of isolation.

Primary factors

History and age

The biodiversity of an ecosystem will depend on the time it has existed and previous history. History as an explanation for today's patterns of biodiversity was the factor that twentieth century ecology forgot. In part this was because history seemed little more than a 'Just So' story; species are where they are because that is where they were, the patterns untestable without the use of a time machine. Modern biodiversity is inevitably contingent on past events. Increased understanding of the Earth's history has sharpened our awareness of history as a factor underlying existing biodiversity.

The distribution of related taxa across widely separated continents, especially in the southern hemisphere, had proven a challenge to nineteenth century biologists. The scatter was explained by lineages jumping across oceans. The revelation of continental drift during the 1960s permitted an alternative view. Distributions today are the result of two historic processes; the origination of lineages on a united super-continent and the subsequent split and continental drift of fragments. This physical splitting is **vicariation** and the resulting diaspora called **vicariance**. Vicariance patterns depend on where the common origin range was, how the now separate fragments once joined, and where splits arose. Large-scale dispersal of lineages, which might include some jump dispersal, is responsible for range expansions, which then might suffer vicariation. In addition to continental drift a gradual expansion of the Earth resulting in a stretched crust may be a historical factor forming vicariant patterns. (See Figure 2.1.)

A second major historical influence is **refugia**, discrete areas acting as sanctuaries during times of global stress. They form the shrunken fragments of larger **biomes** and expand, reforming into wider swathes during benign periods. The idea was developed from the fate of tropical rain forests during the Pleistocene glaciations of the northern hemisphere, though could apply to many global biomes. Not only do the refugia conserve existing biodiversity but the fragmentation may increase net diversity due to speciation, the new species surviving later reunion of refugias. The great age of tropical rainforests has been used as an explanation that seems to contradict the refugia hypothesis, the **stability–time hypothesis**. Tropical rainforests are ancient stable habitats, allowing prolonged accumulation of species. Rainforests have existed for tens of millions of years. Pleistocene glaciations are much more recent. Both historic processes have added to the riot of biodiversity.

Age and history are important. Massive physical processes such as continental drift and glaciation linked to dispersal of lineages and evolution explain the broad patterns of biodiversity distribution and endemism. Large-scale biodiversity differences between continents are contingent on the past. We would not have the unusual faunas of Australia and Madagascar if it were not for their history.

Gradients

Biodiversity does not litter the planet at random but typically changes across environmental gradients. (See Figure 2.2.) The most famous gradient is the increase in biodiversity from high latitudes, the poles, to low latitudes, the tropics. Trends of light, temperature, climate and seasonality facilitate diversity. Increased diversity towards the tropics occurs in land mammals, birds, reptiles, amphibians, insects, snails, clams, trees and plankton. Marine fish diversity increases towards the tropics but some marine trends are sharply stepped, linked to distinct fronts between massive water fronts. Exceptions to the trend include some marine groups, e.g. penguins, perhaps the result of their south polar origins, although the diversity of ecologically similar auks in high northern hemisphere waters hints at other influences. Terrestrial exceptions include conifers which are most diverse in temperate latitudes. The decrease of diversity with altitude is a response to similar environmental conditions, with similar exceptions of conifers most diverse at mid-altitudes of mountain ranges.

Biodiversity changes with depth underwater but the deep sea remains largely unexplored so generalisations are risky. Diversity seems to decrease with depth, certainly in high diversity habitats such as coral reefs. Other marine systems show greatest species richness at intermediate depths. Samples of deep-sea floor life

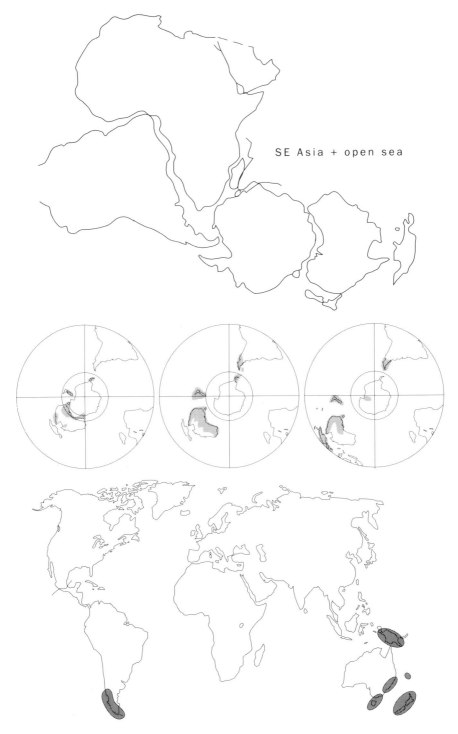

SE Asia + open sea

Figure 2.1 *Distribution of the Southern Beech, genus* Nothofagus, *a classic example of vicariance. (a) The modern continents when joined as Gondwanaland; (b) the subsequent separation of southern continents. The shaded areas show areas with fossil evidence for presence of* Nothofagus; *(c) Modern distribution of* Nothofagus, *apparently very disjunct but in regions historically close together*

Source: Redrawn from Tarling (1972). Another Gondwanaland, *Nature*, 238, 92–93 and Hill (1992) *Nothofagus*: evolution from a southern perspective, *Trends in Ecology and Evolution*, 7, 190–194.

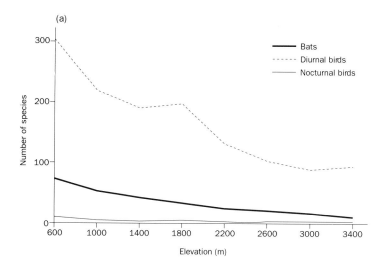

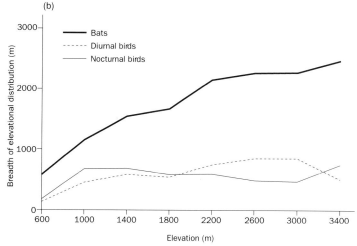

Figure 2.2 *Numbers of bat and bird species recorded over an altitude gradient in the Peruvian Andes. Numbers of species decline with increasing elevation. As altitude increases bat species show wider ranges, perhaps having to roam further to survive, in turn restricting numbers that may live together. Oddly birds do not show this trend, so different taxa may be affected by gradients in different ways*

Source: Redrawn from Graham (1990) Bats versus birds: comparison among Peruvian volant vertebrate faunas along an environmental gradient. *Journal of Biogeography*, 17, 657-668.

increasingly hint at a rich variety. Depth is not simply an underwater analogue of altitude but a response to the synergy of light, temperature, pressure and dissolved substances such as oxygen. Geographical gradients also include aridity and salinity. Diversity is lower in arid regions compared to wet zones of similar area, altitude and distance to the sea. Aquatic diversity is highest in either sea or freshwater but decreases sharply at brackish concentrations in between. (See Plate 7.)

Gradients can be divided between regulation gradients where the controlling factor is not used directly (e.g. altitude) or resource gradients where it is (e.g. nutrients), altering availability. Although biodiversity gradients are so striking the explanation of how they arise is difficult.

Most attention has focused on the polar–tropical gradient. **Time** may be a factor. Pleistocene glaciations smothered polar and temperate regions causing extinctions. The recent retreat of the ice has not allowed time for species to recolonise higher latitudes. Meantime the tropics survived allowing time for speciation. So time as a factor is linked with disturbance. This combination is encapsulated by the idea of ecosystem **stability**. The stability of ecosystems (combining **resistance**, the ability to absorb impacts without suffering change and **resilience**, the ability to recover from changes that are inflicted) was seen as a dominant influence. Stability encouraged complexity, which in turn reinforced stability. These ideas are encapsulated as the stability-time

Plate 7 *A diversity gradient with altitude. Creag Fhiaclach in Scotland is believed to have the only natural tree line in Britain. The Caledonian pine woods of the glens give way to birch then montane herb and grass communities as altitude increases and conditions become increasingly harsh*

hypothesis. Contrarily disruption and instability have been cited as promoters of tropical diversity. Contraction and fragmentation of tropical forests followed by expansion and fusion in response to glacial cycles may have provided opportunity for local speciation in the fragments, the diversification surviving once the forests expanded and coalesced.

Habitat heterogeneity is another ambiguous factor. Tropical forests and coral reefs are more heterogeneous, providing more refuges and opportunities. But rainforests and coral structures are a response to gradients, not a cause, though in turn they provide an arena for the diversification of other life dependent on them.

Tropical diversity seems closely tied to benign conditions, neither too harsh or unpredictable. This link has been termed the **favourableness** of the environment. Favourable is tricky to define, not least because a harsh environment is not harsh for taxa adapted to live there. Tropical forests are warm but not too hot, wet but not waterlogged and not highly seasonal. These generous conditions permit high productivity, high biomass and high diversity. Low diversity occurs in either constant but extreme (e.g. deserts) or fluctuating (e.g. estuarine) habitats.

Favourableness promotes **productivity**. The low diversity of large old predictable (all characteristics supposed to promote diversity) habitats such as deserts suggests restrictions on speciation or colonisation, perhaps due to low productivity. Fundamental constraints such as temperature and light act as severe limitations. High tropical diversity requires that productivity is divided between more species not monopolised by larger populations of the same, few species found elsewhere. A diversity of specialist species may exploit the full range of productivity very efficiently, excluding a more limited variety of generalists. The favourable tropics cover large areas, ensuring total productivity is high, effectively a bigger energy cake to be divided, sufficient to maintain the small populations typical of many rarer species, without undue risk of accidental extinction associated with tiny populations or ranges. This possibility is termed the **species–energy hypothesis**.

The lack of physical constraints in the benign tropics may permit more interactions between species, more intense, more specialised, more likely to reach a conclusion as arms races spiral and competitors outwit one another. The **interaction hypothesis** is awkward to demonstrate and circular: diversity begets diversity. Nonetheless interactions once accelerated in favourable conditions may snowball.

An alternative to this tangle of favourableness, productivity and interaction is that tropical high diversity is nothing more than a variation on the theme of another primary factor, area. The tropics harbour greater diversity because they cover a greater area. Their extent is not only absolutely larger than other global biomes but northern and southern hemisphere tropics join, unlike the separation of temperate or Arctic zones.

Area

Biodiversity increases with increasing area. This rule of thumb remains an awkward ecological pattern to explain. A large area of polar habitat will contain many fewer species than a smaller area of rainforest: glib comparisons can be dangerous. The pattern may be little more than a sampling artefact since bigger areas act much like bigger nets, collecting up more species without any especially ecological mechanisms at work.

The link between area and biodiversity is enshrined in **species–area relationships**. Numbers of species increase with increasing area. Data depicting this pattern are commonly plotted to give a species–area curve, the increase of species with area typically slowing as the pool of species still to find is used. Species number and area are often plotted transformed into their logarithms, resulting in a plot approximating to a straight line rather than a curve. Straight lines are much easier to model as equations and many species–area relationships can be described by the equation $S = cA^z$. S is the number of species and A the area; c and z describe how species numbers change with area; c is the number of species per unit area and then z adjusts the effects of area which may vary between different habitat types. The hunt for consistent patterns has focused on z. Very similar ranges of z, between 0.25–0.3, for different habitats have been cited as evidence supporting some fundamental ecological rule governing species diversity in relation to area but z can be much more varied. (See Figure 2.3.)

Explanations for the effect of area on diversity vary from suggestions that the patterns are little more than statistical artefacts through to area as the predominant motor of ecological diversity. If all species were distributed at random a bigger area would contain more, but since they do not occur at random, except perhaps at a local scale, this simple explanation does not hold. Other non-random distribution patterns may generate species–area curves. Many ecosystems support total numbers of species and populations of each that follow a very particular pattern. Many species occur at low or moderate densities, a few at very low and a very few at very high abundances. Plot the numbers of species with populations in each category and the distribution of species across population categories fits a pattern called the log-normal. Different sized areas drawing their species from a total pool that matches a log-normal pattern can create a classic species–area curve. The important ecological processes are those that produce the log-normal pattern in the first place, not the area.

Larger areas contain a greater variety of habitats, **habitat heterogeneity**. Not only will larger areas contain more types but the types will include increasingly different habitats. Combine the increased variety of habitats as area increases with the

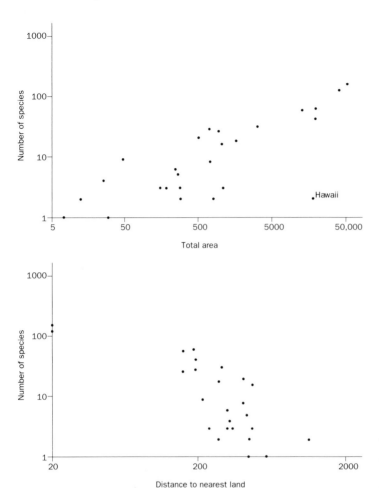

Figure 2.3 *Species:area and species:isolation patterns for butterflies of the Pacific islands. Numbers of species found in diferent archipelagos are plotted against archipelago area and distance from nearest mainland. Species numbers increase with area, decrease with isolation. Note the lack of species in the Hawaiian islands, despite their total size due to extreme isolation. Area and isolation alone can explain 75 per cent of the butterfly species diversity of Pacific islands*

Source: Redrawm from Alder and Dudley (1994) Butterfly biogeography and endemism on tropical Pacific Islands, *Biological Journal of the Linnean Society*, 51, 151–162.

variations in detail at different scales and area's effect in determining species number may be the result of habitat heterogeneity.

The extent of an area will affect evolutionary processes, the opportunity for speciation and risk of extinction. Biodiversity differences between provinces and continents of varying size echo their history. The **species–time hypothesis** recognises the role of time, larger areas harbouring more species due to more opportunity to speciate and less risk of extinction.

Area remains a contentious factor. In an extremely thorough exploration of the ecology of diversity, Rosenzweig entitles a summary section 'A large piece of the puzzle: area effects', combining statistical, habitat heterogeneity and evolutionary processes. Other authors demean area as little more than a handy surrogate for the underlying mechanisms.

Isolation

More isolated habitats harbour a reduced biodiversity. Isolation is not simply the distance between a source of colonists and an island but the interaction of distance, species dispersal mechanisms and any special hazards disrupting access. Rosenzweig distinguishes mainland (essentially area) from island (isolation) patterns. Islands include literal oceanic isles but also separate patches of the same habitat scattered through the landscape. Genuine islands may be self-contained with species originating entirely from immigration and none dependent on continual topping up by immigrants for survival. Mainlands may be populated by taxa that originated from speciation within the area. By this definition Hawaii is a mainland.

Island biogeography, epitomised by MacArthur and Wilson's Theory of Island Biogeography, depends on two processes: immigration which varies with isolation from colonisation source to island and extinction, which depends on island area. As more species colonise the pool of potential colonists is used up. In the meantime extinctions increase, perhaps simply because every species has some chance of accidental extinction or perhaps because of interactions between the increasingly cramped arrivals. An equilibrium diversity is reached once immigration and extinction cancel each other out. This may be a **dynamic equilibrium**, as species colonise, go extinct and recolonise. A **turnover** of species is common though hard to measure since a species may be found, then go extinct and re-establish itself before the next survey, so-called cryptoturnover.

Island biogeography predicts that should an island be perturbed so that it harbours more or less species than the equilibrium and assuming the perturbation does not alter the island habitat, then the equilibrium will re-establish. Experiments following the fate of islands intentionally stripped of fauna suggest that not only does the equilibrium restore but other features such as the general patterns of predator and prey species, the trophic structure, are rebuilt to mirror the pre-disruption status quo. However, the actual species involved may differ. Such results suggest that there are places to be filled, limits and opportunities for diversity.

Isolation and area create **source** and **sink** habitats. Sinks are patches in which a species is unable to maintain a viable population without topping up by immigrants. Sources are areas where a species' reproduction is sufficient for survival. The arrival of immigrants to a sink patch is called the **rescue effect**.

Isolation and inter-patch processes have recently been developed as **metapopulation** ecology. A metapopulation is a set of populations in separate patches but linked by immigration. (See Figure 2.4.)

Ecological templates

Productivity

On a global scale biodiversity increases with rising productivity, one factor influencing planetary biodiversity gradients. At a local scale productivity may limit or promote species diversity. At low productivities diversity is low. Some species simply cannot survive on the limited resources. This may be a particular limitation to specialists. At very high local productivity diversity declines as a few competitively dominant species, freed from resource constraints, monopolise the habitat. Other species are ousted, unable to resist the competition or perhaps simply unable to utilise the riches. The increased nutrients may also promote dynamic instability in **food**

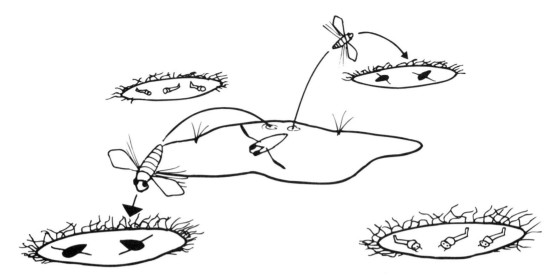

Figure 2.4 *A metapopulation of greater waterboatmen,* Notonecta *sp. The Boatmen survive in the permanent central (source) pond. Peripheral ponds may be colonised by flying adults, which then breed. Ponds not colonised contain prey species that are wiped out in the presence of the Boatmen. The peripheral ponds dry out each year so that the ecological stage is reset. Note how habitat patchiness and disturbance, dispersal and predation all affect local diversity*

webs, the resulting species interactions leading to extinctions as populations fluctuate dangerously. At intermediate productivity levels there are sufficient resources to allow many species to survive but none may take over, the **intermediate productivity hypothesis**. Intermediate productivity may maintain diversity as the variety of life generates local habitat heterogeneity, both of physical structures, e.g. plant growth forms, and of resources such as nutrients or moisture. So productivity links to the resource needs of individual species and **competitive interactions**. Productivity has also been linked to disturbance in the dynamic equilibrium model of species diversity. In this model diversity is determined by the synergy of production (and its effects on population growth and competition) and disturbance (varying in frequency and intensity). (See Plate 8.)

Plate 8 *A species rich marsh, Aberlady Bay, Scotland, a Local Nature Reserve. The marsh nutrient supply is productive enough to support a variety of species but not over-enriched (the fate of many wetlands vulnerable to agricultural fertiliser run-off) which reduces diversity as a few species monopolise the habitat*

Productivity may also tie in with scaleless patterns of species diversity such as **species–abundance patterns**. If species diversity fits species–abundance patterns, which are not purely ecological but much more general rules governing collections of things, and the productivity pie is sliced up accordingly then the more pie (the greater the productivity), the more species will get a share, so occur in a habitat.

Local declines in diversity at increased productivity have been shown in terrestrial and aquatic habitats, especially where nutrient enrichment occurs. Enriched aquatic systems lose diversity as algae, able to respond quickly to increased nutrients, bloom and smother larger plants. Enriched grassland swards also lose species as a few coarse grass species take over. In both cases removal of nutrients can reverse the trend and local diversity recovers.

Scaleless patterns

Scaleless patterns include apparent correlations between bodysize, population abundance and number of species, plus rules for the structure of food webs.

The relationships between species numbers, population abundance and body size have been a focus of recent work. Plot the numbers of species in a community against their populations and some consistent species–abundance patterns emerge. Models to explain these patterns focus on how species divide up available resources (resource apportionment) or statistical patterns that apply to collections of things in general, ecological or otherwise. For example, the 'geometric series' assumes that a first species to establish monopolises a share of the resources, the second species in takes the same percentage of the remaining resources, the third the same percentage of what is left and so on until all resources are used up. Other models are having no specifically ecological roots and making no assumptions about what the species do. Such models can just as well describe patterns of diversity in collections of any objects, insects, kitchen cutlery or chocolate bars. For example, the log-normal distribution pattern arises when the abundance of items (in this case populations of species) is the result of many random independent influences.

Species–size

Species–size patterns take a similar approach, plotting the number of species against categories of body size, looking for consistent patterns and trying to explain these with general models.

Species–abundance–size

Species–abundance–size patterns have proved frustrating. There are some potentially important lessons for biodiversity and conservation, in particular if any consistent patterns linking abundance to size or diversity suggest that there are some population sizes simply too small to survive. It is possible to find ecological communities to fit each model but none are generally applicable.

Food-webs

Consistent patterns in food webs (species linked as those higher up the web exploit those below for food) suggest rules govern web structure. A species' position in a web is defined by how it captures energy. Primary producers trap the raw energy of sunlight, sometimes other sources, and convert this into stored chemical energy. Primary consumers including herbivores eat the primary producers and secondary consumers, predators and parasites, prey on the primary consumers. Since predators have their own enemies there may be several levels of secondary consumers. Each of these categories is a trophic level.

Food webs show consistent patterns. First most contain no more than three or four levels. Longer webs can be found, more out of a love of the unusual. For example, some marine webs may go up to eight, assuming tortuous series of links from tiniest plankton to largest sharks, by way of four or five levels of secondary consumers eating one another. The consistent rule stays sound. Most food webs reach three or four trophic levels. The second pattern is that omnivores, defined as a species that takes at least 5 per cent of its food from another trophic level than its main supply, are rare. Third ratios of numbers of predator to victim species, links per species, and links between different levels is also strikingly consistent in real webs. As diversity of one increases, so does diversity of the other.

Limits to food web length were originally ascribed to **energy transfer** from one level to the next. Only 10 to 20 per cent of energy eaten is assimilated, the bulk being lost via excretion or metabolism, so available energy rapidly peters out up the chain, leaving too little to support any more levels. If available energy was the limiting factor webs from more productive habitats should be longer, but they are not. Only in the most unproductive habitats, e.g. terrestrial Arctic, is there evidence for truncation due to low energy. An alternative to energy limits is **dynamic stability** of webs as they become ever longer. Model food webs can be built to mimic realistic producer, predator and prey interactions, the fate of populations tracked over many generations. The longer the model food webs the more unstable they become. As populations rise and fall interactions ripple through the web and feedback lurches through the system. Long webs lose levels, collapsing back to more stable two, three or four levels. Dynamic models generate realistic web length limits, realistic predator–victim ratios, numbers of links and rare omnivory.

Energy flow and dynamics may be linked. In low productivity systems stability increases with an increase in productivity. In high productivity systems any further increase in productivity promotes instability. Taken together productivity may limit web length in low productivity habitats, instability in high productivity systems.

Realistic links and ratios can be created by a third, strikingly simple, theory, the **Cascade model**. Ignoring population dynamics and energy the Cascade model suggests that species are arranged in a hierarchy, perhaps by body size, so that each only feeds on those below and is fed on by those above. The total number of links per species is fixed, but how species link is randomised. The Cascade model predicts web length limits poorly.

Habitat heterogeneity

A varied physical environment harbours a greater biodiversity. Heterogeneity refers to the kinds, variety and distribution of different types of habitat. All environmental factors vary across time and space. The more variation, of size, age, type and amount of patch, the greater the diversity of life. Patches can be physically discrete areas or variations in the distribution of resources. This patchiness, coupled with changes over time, mitigates against a few species monopolising the environment. No species can be good at exploiting every environment. There is a trade-off, strong evolutionary pressures to optimise strengths while losing less successful attributes. The process will be self-reinforcing, with less and less to be gained from trying to exploit patches that other species can utilise more effectively. Even an apparently homogeneous environment will become varied as different species colonise, altering their local environment. The results is that different patches will vary in their usefulness to species. (See Figure 2.5.)

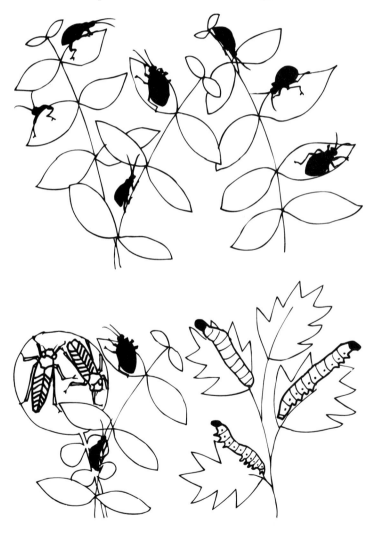

Patchiness mitigates against monopoly by just a few species. Changes in one patch (losing out to a competitor) may be compensated for by a different fate in another patch (the victorious competitor wiped out by its predator). Local variation between patches results in regional coexistence of species. This is especially so if the ecology of separate patches is out of sync. For example, if disease strikes at a particular stage in a life history populations will survive in some patches that are not at that vulnerable stage. Habitat heterogeneity has been shown to promote co-existence of predators

Figure 2.5 *Habitat heterogeneity. The greater the variety of habitats, in this case plant species, the greater the variety of species, in this case insects associated with each plant, although the overall quantity of habitat has not changed*

and prey, e.g. many insect predators such as ladybirds can wipe out an individual patch of prey but cannot find all prey colonies hidden in different patches, and of competitors, e.g. many plant species which differ so little in requirements that competition would be severe co-exist regionally scattered across separate patches. So the fate of individual patches links with **isolation** and **area** to create regional diversity. Temporal variation will also maintain diversity. Seasonal changes can upset the otherwise inevitable outcomes of interactions between species, resetting the ecological stage. Different species may suffer resource bottlenecks at different times of year, even if they are using the same resource. (See Plate 9.)

Plate 9 *Habitat heterogeneity. This small stretch of river supports discrete patches of vegetation, submerged and emergent as well as open water. The different patches each provide different habitat and alter the physical environment, increasing the local diversity of animals*

Habitat diversity (heterogeneity and complexity) and species diversity co-evolve. The richness of habitats in a region depends in part on the species to recognise their existence. Species discriminate the finer grain of habitats (be it patches or complexity) because selection forces them to. Natural selection finds the habitat diversity. Some regions with apparently very limited environmental diversity, e.g. the Fynbos of South Africa or Rift Valley Lakes, can support a surprising diversity of species (plants and fish respectively) that show a very fine-grained division of the environment.

Habitat complexity

The more complex the architecture of a habitat the greater the diversity of species. Complexity refers to the structural architecture of each habitat. The amount and shape of structural complexity in the environment will create opportunities for species. A structurally complex ecosystem (e.g. the 3-D architecture of a forest) provides more physical habitats for species than a simple one (e.g. an equivalent area of tidal mud). Habitat complexity sounds intuitively obvious but is difficult to untangle from other factors. For example, in the forest/mudflat comparison above the diversity of plants in an area of wood provides many more sources of food for herbivores (and in turn their enemies) than a mudflat. (See Figure 2.6.)

Figure 2.6 *The greater the habitat complexity the greater the diversity of species. In this case the physically complex plant hosts more species. The variety of different sized and shaped spaces on the complex plant creates opportunities for more species, which vary in body size*

Plate 10 *Habitat complexity. The simple, straight leaves of the bur reed,* **Sparganium erectum,** *provide a less complex habitat than the branches of water mint,* **Mentha aquatica.** *More species can exploit the greater complexity offered by the mint habitat*

An important insight into complexity is that there is more of it at smaller scales. Many natural phenomena, whether soil structures or vegetation, are fractal. They do not have a finite size but measurements of dimensions such as area or edge increase the finer the scale at which they are measured. To a 1m tall grazing deer a small herb like a buttercup is scarcely a mouthful; to a 5 mm long insect a buttercup is a complex climbing frame, with leaves, shoots, flowers and buds, each offering different resources. So the availability of microhabitats increases at smaller and smaller scales. As a result species biodiversity increases at smaller scales. There is more room for organisms of smaller body sizes. The biodiversity of terrestrial insects may in part be the result of this pattern, linked to the complexity and scale of the physical habitat that plants create. Many habitats are exactly the right scale for the diversification of insect-sized life. (See Plate 10.)

Intimate processes

Succession

Diversity changes as species establish, interact and alter the environment. The biodiversity of a site increases over ecological time with immigration, establishment and loss of species creating a successional pattern. Succession is the non-seasonal turnover of species over time. The turnover is predictable with recognisable early colonists and late survivors. The progress of succession is directional, the sequence of early and late successional communities repeated across separate patches of the same habitat type in the locale. The directional replicated changes may converge on a final community. Successional change comprises a sense of direction, of species turnover and of physical change as the biodiversity creates its own structures and alters the environment.

Early twentieth century ecology was mesmerised by the hint of rules and progress hinted at in successional patterns. Theories in successional change were developed primarily for plant communities. The sequence of plants is a major factor in creating landscape, so succession may be important in determining plant biodiversity and, in turn, the diversity of wildlife exploiting the plants and the physical habitat they provide. Original theories saw successional change as a predictable drive towards a final ecological superorganism, the species of plants so integrated that characteristics of each community were greater than the sum of the parts, with the result an inevitable **climax community**, perhaps fine tuned to maximise efficient energy transfer. This vision faltered when faced with successions leading to multiple outcomes, none conspicuously a climax, successions that appeared to reverse and no evidence for increased energy efficiency.

Succession is now widely regarded as the result of individual species' life histories and attributes, each living its own life and the consistent patterns arising because of the same main environmental pressures acting on the same basic set of life forms. The attributes of plants, interactions with and changes to their environment and interactions with herbivores and diseases that exploit the plants in turn drive the process. Succession arises from the intimate ecology of physiology and life histories of species interacting with their local environment. Five processes interact: disturbance that creates an opening, immigration, establishment, competition and reaction as species alter their habitat. Important species attributes include dynamics (maximum growth rates, size, longevity and establishment) and physiology (requirements for or tolerance of light, water and nutrients).

Succession is a dynamic process and diversity can increase or decrease. Succession on virgin habitat such as a volcanic island or newly exposed glacial morraine is called primary succession. Secondary succession occurs on terrain stripped of plants but retaining previous biological influences such as soil conditioning and seed banks. Succession can be categorised by the driving forces of change. Change in response to external influences, e.g. climate change, is allogenic. If driven by the organisms themselves, altering their habitat succession is autogenic. Degradative succession occurs on organic debris such as a cow pat or dead leaf, the end result being the habitat's destruction.

Succession may be yet another ecological process that is nothing more than a statistical phenomenon that just happens to involve wildlife. Successional patterns can be modelled using nothing more than the probabilities that a species will arrive or be lost, regardless of the previous history of the site or present inhabitants. Species come and go at random.

Huston comments, 'Succession is the key to understanding the regulation of nearly all aspects of biodiversity or ecological time series.' While Rosenzweig frets, 'We have to ask the plant ecologists to return to their succession data . . . and tell us whether a consistent pattern exists.'

Interactions between species

Interactions between species can promote or destroy local diversity in the short term but are a major force for evolutionary diversification. Interactions between species fall into four broad categories.

1 **Competition**. Species contest a resource. Where the resource is limited inferior competitors may be driven to extinction. Competition is often depicted as detrimental to all species involved, but is often asymmetrical, with one suffering, the other unaffected.

2 **Exploitative**. One species exploits another as a food source. Exploitative interactions include predation, parasitism, disease and herbivory. The victim loses, the exploiter gains. Diversity is decreased if exploiters wipe out victims or promoted if the monopoly of a few dominant competitors is prevented.

3 **Mutualisms**. Both species benefit from the presence of the other and their interaction. Such relationships are often tightly forged, by coevolution, into a mutual dependency. Mutualisms promote diversity by opening up new ecological opportunities.

4 **Engineering**. Interactions due to the activities of one species benefiting others, but with no reciprocal gain and typically not tightly coupled by coevolution. Ecological engineering may be physcial, e.g. Beaver dams, or functional, e.g. nutrient cycling.

Competition between species occurs when they use a resource in such short supply that there is not enough to sustain their potential reproduction, growth, abundance or distribution. Competition was regarded as the major process regulating local diversity. Species exactly overlapping in their exploitation of a resource could not coexist indefinitely. One wins, others lose. Competition could exclude species, lowering local diversity. However, combined with disturbance and patchiness, a mosaic of competitive interactions could result in a variety of outcomes, maintaining diversity. Longer term evolutionary pressures would select for species adapted to avoid competition by niche differentiation (living in a sufficiently different way to avoid competition), character displacement (anatomical differences), niche shifts (short-term modification of niche to avoid competition) and complementarity (species sharing a resource would use it at different times or places). A species cannot be good at everything and there is a strong trade-off between improvement of core abilities and evolutionary abandonment of what is done poorly. Competition provides a selection pressure with species diversifying by refinement of different abilities simply to stay in the evolutionary game. Competition theory suggested that there were limits to how similar species could be and therefore a limit to how many species could pack into an ecosystem but golden rules proved elusive. Existing niche differentiation might show competition had been an important evolutionary pressure in the past but was not active today, the species and diversity patterns ghosts of competition past. Re-evaluation of competition's role has recognised its importance but as one of several interactions. (See Figure 2.7.)

Competition in evolutionary time is also a mixed force. Avoidance of competition by differentiation, complementary rather than overlapping exploitation of resources and exploitation of entirely new resources are powerful forces to speciation. However, the eclipse of some higher taxonomic groupings, e.g. Palaeozoic squid ousted as dominant marine predators by the rise of the fish, represents a loss of diversity.

Competition's fall from pre-eminence as the process determining local diversity was in part due to increased understanding of the importance of **exploitative interactions**, especially predation. Predation and competition were often cited as if mutually exclusive rivals but they have very similar effects. Exploitation can increase or decrease local diversity and is a force for evolutionary diversification. The difference

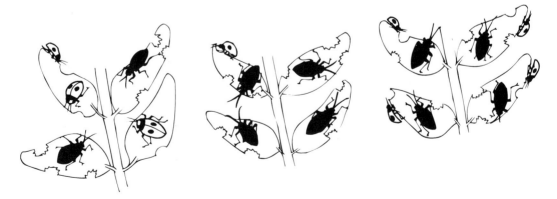

Figure 2.7 *Competition. Two beetle species feed on the same host plant. Competition almost wipes out one in the short term, threatening local diversity, but evolutionary diversification due to separation of feeding sites consolidates diversity in the longer term*

lies in the structure of the interaction. Competition for food involves species feeding at the same trophic level, exploitative interactions involve links between different levels (eating your competitor could count as either). The impact of exploiters varies with the intensity of interaction, their behaviours and synergy with competitive processes. Generalist predators maintain diversity by eating anything they find in proportion to its availability or homing in on the most abundant prey by switching their hunting activity to concentrate on it and massing where the prey is concentrated. Either strategy inflicts most damage on prey species that potentially monopolise a habitat by outcompeting other species. Specialist predators may have the same effect if they specialise on would-be monopolisers but can reduce diversity if their target is a rare taxa. Exploitation works like physical disturbance. Too little allows a few species to take over, too much reduces all prey. Intermediate levels generally promote local diversity (the **intermediate predation hypothesis**). The role of predators prompted the idea of **keystone species**, which by their presence or absence define local community diversity. The keystone idea is now extended to include other roles such as keystone engineers or mutualists.

Natural historians recognised the importance of **mutualisms**, perhaps as evidence of a divinely created harmonious nature, until overthrown by post-Darwinian evolutionary theory with its imagery of contest and struggle. Mutualism became the forgotten triplet of competition and exploitation but the importance of mutually beneficial links is fundamental to ecological diversity and planetary health. Mutualisms are defined by benefit to both species, though the benefits could be seen as reciprocal selfishness rather than benign altruism. Mutualisms are typically cemented by specific behavioural or physical ties forged by coevolution between the participants. These ties can be so specialised that neither partner can survive without the other and no alternative species can step in should one partner become extinct. Mutualisms fall into four categories. **Trophic mutualisms** are built on the exchange of food. Most often the two partners provide nutrients that the other cannot obtain or in sufficient quantities. Examples include coral polyp/algae, forest trees/fungal mycorrhizae and plants/nitrogen fixing bacteria. **Pollination mutualisms** and **seed dispersal mutualisms** involve plants and animals. In return for bringing pollen to a flower or transporting away seeds the animals are rewarded with food such as nectar or pollen. **Protection mutualisms** comprise interactions in which at least one partner benefits from reduced exploitation or competition, e.g. ants tending aphids repel parasites and predators, in return receiving honey dew; cleaner fish ridding other fish of parasites, a food reward for the cleaner.

The most biodiverse ecosystems on Earth, tropical rainforests and coral reefs, are founded on trophic mutualisms. Mutualisms open up new niches. Even species without direct mutualists have indirect benefactors. In a food chain linking plant, herbivore, predator and parasite the plant and predator are mutualists, the plant hosting the predators food the predator, preventing its prey from overexploiting the plant. Similarly the herbivore and predator's parasite are mutualists. Mutualists are an additional resource prompting competition as well as cheat strategies. The image of nature conjured up by the importance and variety of mutualisms has been described as green in root and flower.

The importance of **ecological engineers** is a recent insight, born of the increasing integration of ecosystem ecology (concerned with energy and material fluxes) and community ecology (founded on species). Living species drive abiotic processes and engineer their environment. The activities of some ecological engineers could be described as mutualistic but they are different because of the lack of close evolutionary ties. In a classic example of engineering the beaver, constructing a pond for its own benefit, inadvertently provides a habitat for many other species. The beaver's actions are vital but not coevolved with the beneficiaries. Engineering therefore includes many interactions where one participant benefits, the other is unaffected. Engineers' impacts work either on biogeochemical cycles or the habitat structure. Some **structural engineers** create physical structure of their own bodies, e.g. coral and trees, so-called **autogenic engineering**. Others create later existing structures, e.g. beavers and woodpeckers, so-called **allogenic engineering**. **Biogeochemical engineers** alter chemical cycles and nutrient availability, e.g. bioturbation of sediments altering nutrient (resource) and oxygen (regulation) gradients.

Reawakening interest in mutualisms and the importance of engineers driving ecosystem function have altered our understanding of the role of species interactions as a source of biodiversity. Competitive and exploitative interactions are very important, creating local variations and evolutionary escalation. Besides the direct impact of exploiters on victims and competitors on each other indirect effects ripple through ecosystems. Many competitors or victims may escape single strong interactions but are buffeted by many diffuse conflicts. The parasites of a herbivore are the plants' mutualists. Two competitors that share a predator magnify predation on each other since their enemy has a larger total pool of prey. Two competitors with different predators benefit from the activities of each other's enemies.

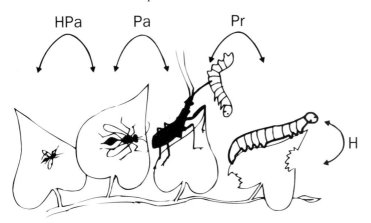

HPa **Pa** **Pr** **H**

Figure 2.8 *Trophic interactions. Herbivory, H; predation, Pr; parasitism, Pa; hyper-parasitism (parasite of a parasite), Hpa. In addition the assassin bug predator of the caterpillar is a mutualist of the plant, killing herbivores. The hyper-parasite a mutualist of the assassin bug, killing parasites so restricting their future impact on the bug, and therefore is also a mutualist of the plant*

Meanwhile ecological engineers riddle their habitats with geochemical variation and physical structure. Evolution multiplies these intimate links, particularly host–parasite systems as host speciation spawns parasite co-speciation. This intimate tangle, populated by mutualists and driven by ecological engineers, has challenged the mechanical reductionist ecology of biodiversity. Ecosystems may be examples of **complex adaptive systems**, defined by their diversity of components, complexity derived from simple rules generating varied patterns, with energy and materials in constant flux and the ability to adapt and evolve. The sense of patterns emerging that are more than or at least different to the sum of their parts shows the wheel of scientific fortune changing. Species interactions are vital to the generation and maintenance of biodiversity. So biodiversity begets biodiversity after all. (See Figure 2.8.)

Disturbance

Natural disturbance is an important factor in the creation or destruction of local diversity. Disturbance takes many forms and acts at many scales. As an intimate ecological factor disturbance includes physical abiotic impacts such as fire, flood, storm, landslide and treefall. Disturbance is defined as a discrete event, a sudden physical destruction, causing mortality faster than populations or cover can be replaced. The loss frees up resources which can be exploited by survivors and colonists. Disturbance is different from continuous stress which slows growth or rare massive catastrophes, e.g. meteor impacts. Local physical disturbance is the ecological process outlined here, though predation and disease are also a form of disturbance, but the result of biotic interactions.

The impact of any disruption varies with five factors: disturbance frequency (number and predictability); intensity; synchrony (one large disturbance versus many smaller out of sync patches, plus timing relative to victims' life histories); size of gap that results; and effects on resources. The disturbance alters biodiversity by immediate deaths, changes to resource availability (amount and type), physical alteration of the habitat and evolutionary adaptation.

The effect of disturbance is inextricably linked to successional competitive interactions and habitat patchiness. Disturbance disrupts inevitable outcomes, particularly monopolisation of habitat and resources by a few superior competitors. The disturbance may keep populations so low that resources are never in short supply and competition does not occur. Or disturbance can reset and redirect the struggle, creating diverse outcomes, especially if separate patches are perturbed at different times and intensities, resulting in a mosaic of communities. The disturbance may alter the environment of a patch so much that it is no longer hospitable for previous occupants. Disturbance plays a vital role in conspiring with deterministic forces such as competition, and physical structure such as heterogeneity, to create dynamic patterns of biodiversity. (See Plate 11.)

Disturbance can also decrease biodiversity, if too frequent, severe or extensive to cause extinctions and inhibit recovery. Diversity is greatest at intermediate levels of disturbance, a general result encapsulated in the **intermediate disturbance hypothesis**. Too much disturbance wipes out species, too little allows monopoly. Defining intermediate is awkward and dangerously circular. Intermediate disturbance promotes diversity because it is a level of disturbance at which diversity is maximised. It is better to consider disturbance and competitive population growth rate processes

Plate 11 *Disturbance and succession. Brown* Fucus *seaweeds dominate the middle levels of this rocky shore but patches have been ripped out by disturbance from waves. These openings have been colonised by fast growing green algae so that local diversity increases. A succession of seaweeds will invade, determined by individual attributes of colonisation, competitive ability and resistance to grazers*

simultaneously. The two have been formally linked in the **dynamic equilibrium model** of local diversity. The link highlights intermediate disturbance promoting diversity when growth rates and competition are high. High growth rates as a result of productivity promote diversity by aiding recovery from disturbance. The model suggests disturbance effects vary with trophic level. High level taxa have smaller populations, which are more vulnerable to extinction and require greater energy to recover. An increase in disturbance will endanger higher level diversity more than that of lower trophic levels, especially when productivity is low. At high productivities but low disturbance biotic impacts of keystone predators and herbivores become important as a means of breaking up low diversity monopolies. At low productivities but high disturbance keystone resource species accelerating resource supply are important. At low productivities and disturbance keystone mutualists are important. (See Figure 2.9.)

Dispersal

Diversity varies with species' abilities to reach a site which in turn depends on attributes and behaviour. Dispersal, the movement or transport of individuals, can be very conspicuous whether dandelion seeds on a summer's breeze or herds of migrating wildebeeste. Because dispersal is awkward to measure or model and the fate of those that do not complete their journey often unknown its role was poorly researched. Biodiversity deals in variety, populations, communities, ecosystems, but it is individuals that disperse. Individual based models, taking into account that all individuals and their fates are different, have advanced our understanding of dispersal's consequences for local diversity. At the opposite extreme, large-scale supply side ecology, the rate and timing of recruitment, particularly of dispersing juvenile stages of marine life swirling in gigantic oceanic currents, helps to explain regional and local diversity.

Dispersal is commonly divided between **migration**, the active, mass movements of whole populations and **dispersal**, the active or passive transport of individuals. The individuals in question may be adults, juveniles, eggs, spores or bits capable of vegetative growth. The precise mechanisms, timing and rates are very species specific.

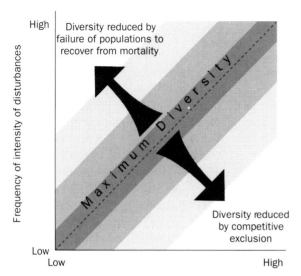

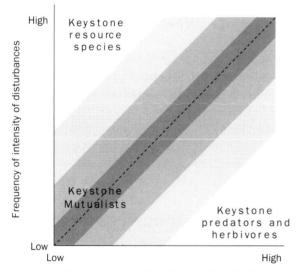

Figure 2.9 *Huston's 1994 synthesis of disturbance, population growth and interactions as determinants of local diversity. The top graph shows diversity reduced if either growth and competition or disturbance is high, without any compensating influence. The graph below shows the importance of different keystone interactions under three disturbance/population growth regimes*

Source: Redrawn from Huston 1994.

Marine plankton may disperse at the mercy of the tides. Some plant seeds disperse in tightly coevolved mutualisms with one species of animal carrier. All species have some power of dispersal, even those in uniform stable habitats. Different types of dispersal create different patterns. **Jump dispersal** over great distances and barriers may found regional populations. **Metapopulation** species rely on dispersal to maintain their scatter of discrete colonies. **Diffusion**, the gradual movement and expansion of range, will mesh with habitat heterogeneity, resulting in an unevenly spread population. Even in a homogeneous terrain diffusion can create patterns. The spread of populations, oscillating through peaks and troughs of abundance typical of many species, will meet like waves on the surface of water. As high and low populations of interacting species ripple across the landscape complex patterns emerge.

Supply side ecology, the rate and pattern of arrival, can be a dominant influence. Inconsistencies in the diversity of many coral reefs and marine benthic communities has shown that variable recruitment, driven by the rate and timing of dispersal, causes local populations occasionally to be deluged by arrivals, while other areas receive none. These occasional events swamp local interactions that regulate diversity in between. There are lessons for conservation. The plagues of the crown of thorns starfish, *Ancanthaster planci*, famed for the havoc they wreak on corals, may outbreak due to occasional unpredictable mass arrival of larvae.

Ultimately the dispersal of individuals builds the ranges within which work evolutionary events such as speciation and ecological primary factors such as vicariation. From time and space to individuals and back again, the ecological factors creating biodiversity are an integrated loop rather than a reductionist shopping list.

The evolution of biodiversity

Ecology describes patterns of and provides explanations for the biodiversity of extant ecosystems. Ecological processes also have evolutionary consequences. They interact with genetic diversity via adaptation, microevolution and speciation. The day-to-day dramas played out on the ecological stage are the driving force for the multiplication of species in evolutionary time. The environment provides continual pressures to diversify via adaptation, innovation and exploitation of new ways of life. Taxonomic biodiversity is the most conspicuous result of ecological and evolutionary processes driving the multiplication of species.

The multiplication of species

Diversification of genes: environmental forces

New species arise from two sources. An individual lineage may change over time so that subsequent generations are sufficiently different to ancestral types that they are classified as a separate species. This is **anagenesis** or **chronospeciation**. There is no net gain of species. Speciation that spawns a net gain in numbers is **cladogenesis**, whether it is the splitting of the ancestral line into two (or more) new species or the budding of new species from an ancestral lineage which also survives. These processes can be gradual, the accumulation of inherited genetic differences selected by ecological processes, or sudden perhaps random with no clear adaptational benefit, Raup's survival of the luckiest.

Taxonomic biodiversity is increased by cladogenesis so it is important to understand why lineages speciate. The genetic variation in populations that underpins all of biodiversity and is central to evolutionary processes is almost unavoidable. Genetic diversity within a species is created both by the environment, as local populations adapt to variations between local conditions, and by processes within the genome that would occur even in a homogeneous environment, e.g. mutation. The drive to genetic diversity is irresistible.

Some genetic diversity can be categorised as **adaptational**. Genetic differences reflect variations in local conditions with selection of **genotypes** beneficial in the prevailing conditions and loss of the sub-optimal. The genetic differences may reflect tolerances to different conditions or create a functional diversity so that individuals in a population can exploit varied opportunities but perfect adaptation is probably impossible, not least because environments constantly alter. Adaptational diversity may create **key innovations** allowing exploitation of lifestyles new to that taxa or novel to any life. Key innovations are one evolutionary factor unleashing adaptive radiations, resulting in geologically rapid diversification of a lineage. Innovations may be utterly new. They also include **preadaptations**, a character allowing ecological or evolutionary exploitation of a new lifestyle while essentially carrying out its original function and **exaptations**, characteristics, again allowing ecological or evolutionary expansion, but the character is used in a different way to its original purpose, fortuitously effective in a new role.

Major periods of innovation, defined by their global extent, diversity of innovations and new ecosystems may spring from revolutions in global biogeochemical cycles. The two main innovation revolutions since the Cambrian (one the Cambrian marine diversification, the other the Mesozoic marine modernisation) coincide with massive biosphere changes such as volcanism, warming and sea level changes which could alter nutrient supply. Increased supplies of nutrients and energy opened up opportunities and once evolution and speciation accelerated a positive feedback of arms races, competition and mutualisms would snowball.

Genetic and resultant evolutionary biodiversity can also arise without any adaptational context. The exchange of genetic material through an interbreeding population results in **genetic drift**. Genetic drift can be a potent force in small populations where the original genetic diversity may be limited to start with, a biased sample of the total variety in a larger population. In a small population a few matings can result in drift rippling through subsequent generations. Small populations prone to such effects can be created in many ways: **founder populations** of a few individuals colonising new habitat; isolated remnants marooned in **refugia**; **bottlenecks** created when populations crash; **metapopulations** where the total population may be high but is scattered between separate patches that do not readily interbreed.

Diversification of genes: genetic systems

Just as the environment forges genetic variation within populations so processes within the genome itself create and maintain diversity. **Mutations** are any changes in the genome be it one nucleotide or larger sequences. Types of mutation are unpredictable, uncontrolled and independent of the natural environment, though rates may vary. Mutations include errors in **DNA** replication, insertion of transposed sections and rearrangement of chromosomes via fusions and inversions. The rate of changes to Eukaryote nucleotides has been estimated as 4×10^{-9} per base pair per generation. Since eukaryotes typically harbour billions of base pairs every individual carries some mutations. We are all mutants. So mutation will create genetic variation even within the least genetically diverse taxon, a source of hope for genetic recovery for some endangered species. Genetic drift creates diversity within a freely mating population. Although the frequency with which individual **alleles** are transmitted to offspring may be broadly predictable (based on their frequency in the parent population and the rules which govern their sharing out between gametes), only a minute fraction of gametes actually contribute to the next generation at fertilisation. The alleles in the billions that do not make it are winnowed. The impact of drift depends on population size and is magnified in small populations. Drift effects can be substantial in large populations, e.g. species with breeding opportunities monopolised by a few individuals such as Red Deer stags coveting harems. The limited number of individuals contributing to the genetic diversity creates a small **effective population size**.

The distribution of individual alleles on chromosomes can also be rearranged by **recombination**. Eukaryote chromosomes exchange sections so that alleles that would otherwise stay linked together are combined into new sets. The total genetic diversity at the allele level does not increase but the permutations do. Recombination is most effective in outbreeding populations. Self-fertile taxa end up recombining with themselves. Prokaryotes have mechanisms that achieve broadly similar remixing of alleles.

Selection and maintenance of genetic diversity

Environmental and genome processes combine to produce genetic diversity within populations. The genetic foundation of each individual is tested by **selection**. Selection occurs within ecological time and forges a lineage through evolutionary time.

Selection acts directly on the **phenotype**. The **genotype** is affected indirectly. **Directional selection** favours individuals with an advantageous characteristic, selecting the genotype responsible and reducing genetic diversity by elimination of alternatives. Since many general phenotype characteristics are a synergy of complementary features not governed by one gene, directional selection effectively selects sets of genes. **Stabilising selection** favours intermediate characteristics and extremes are winnowed out. Stabilising selection fixes combinations of alleles that build the intermediate character rather than variations within each allele, so again genetic diversity is lost. **Disruptive selection** favours different extremes, the intermediates are at a disadvantage and maintains the variation in alleles that cause the switch between extremes, not different alleles themselves. So selection tends to decrease genetic diversity. However, fluctuations through space and time alter selection pressures, even reversing them, providing an important brake to such losses. **Genetic dispersal** and **gene flow** act to homogenise genetic diversity within a population. Highly mobile species show less variation than the more sedentary. The dispersal characteristics of a species are important. Some species known to be poor dispersers nonetheless show little genetic variation throughout their range. This may be because they have only recently expanded into the range, the homogeneity an echo of their close-knit origins. Species living as a metapopulation can show marked losses of genome diversity if extinction of patches wipes out unique variations. The toll from selection and drift is countered by imports from flow. The outcome teeters, critically dependent on how far individuals move between birth and mating versus the strength of selection. There can be a critical distance below which flow can compensate for any effects of selection. The conservation of small fragmented populations often depends on understanding such pressures. In summary, mutation creates genetic diversity, selection and drift destroy.

Breeding systems affect genetic diversity. Inbreeding allows disadvantageous, often recessive alleles expression. Enfeebled offspring show **inbreeding depression**. If they breed in turn the degradation may snowball into a genetic meltdown. This is a particular threat if a once large population is suddenly reduced in size. Genetic meltdown becomes the final straw after the first strike that reduced abundance. Small founder populations which may be **homozygous** to start with so that any deleterious alleles have been weeded out already are in less danger. **Outbreeding** generally suppresses the worst impacts of disadvantageous alleles. Outbreeding depression can occur via immigration of individuals poorly adapted to local conditions which mate with well-adapted locals. There is a balance. Mate with those too close and risk inbreeding, with those from too far and risk outbreeding depression.

Given the environmental and genome promotion of genetic diversity it is not surprising that there is no evidence of limits to local, small-scale selection for want of genetic variation. We do not know if there are limits to major evolutionary changes, for instance, radical rearrangement of body plans. These may be prevented by lack of genetic foundation. Perhaps the genes do exist but developmental rules halt any such revolutions. Even without the capacity for unfettered creativity the environment and genes have conspired to create a wealth of species.

Speciation

Speciation is the creation of new species. While commonly pictured as a diversification or radiation of a lineage, speciation requires mechanisms both to create and then maintain the diversification into cohesive species. Only then can a genotype establish, its individuals isolated from interbreeding with other species. The familiar images of radiating, diversifying lines might be better thought of as collapsed spokes from a wider fan of individuals that once freely interbred.

The threshold from variation within a species (adaptation, microevolution) to speciation and macroevolution is a barrier. The very same processes creating genetic diversity within a species also maintain cohesion. Gene flow may smother variation; unusual genotypes may be less fertile or selected out by the environment or ignored by would-be mates. Selection may hone well-adapted species, marooned on peaks of near perfection, from which any evolutionary departure is difficult.

Mechanisms are needed to push genetic diversity over the speciation threshold. There are four main modes of speciation.

1 Allopatric. Geographical separation of a previously continuous population.

2 Peripatric. A variant of allopatric involving isolation of a peripheral population.

3 Parapatric. Speciation in adjacent populations stretched over an environmental gradient.

4 Sympatric. No separation during speciation.

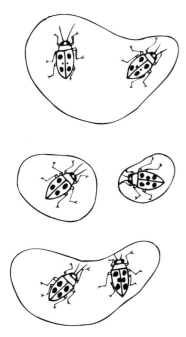

Figure 2.10 *Allopatric speciation. A united population is split, in this case by a physical barrier of rising sea levels. The evolutionary history of the isolated populations differ, to the extent that once reunited they no longer successfully interbreed. Two new species exist where there had been one*

Allopatric speciation

Allopatric speciation requires the separation of populations either by creation of a physical barrier, extinction of the middle range of a population or jump dispersal across an existing barrier prior to speciation. The barrier prevents gene flow and the populations' genomes diverge. The divergence can be adaptive if environments differ either side of the barrier or random due to mutations, recombinations and chance losses. If the diverging populations rejoin they may still interbreed and coalesce, maintain the divergence but with a hybrid zone in between, or not interbreed at all, the latter representing full speciation. This reproductive isolation can be maintained by genetic incompatibility, discrimination in choice of mate or the physical impossibility of mating. (See Figure 2.10.)

Peripatric speciation

Peripatric speciation involves isolation of a small peripheral population from the main range. Otherwise the process of geographical isolation followed by divergence is the same as allopatric speciation. The small size of peripheral populations creates specific features. Small numbers of founders inevitably contain only a limited, biased subset of the original species' gene pool. This provides a head start to divergence, magnified by inbreeding and homozygosity allowing expression of unusual characters previously suppressed. The cause of isolation may add impetus. Founders colonising an island may have abundant resources and little competition. Alternatively peripatric speciation in collapsed remnants of larger populations can face genetic bottlenecks. These **founder effects** emphasise the vagaries but importance of **chance events** and genetic loss. Alternatively adaptations may flourish in new or empty niches. Old genetic complexes are shaken up and innovations and radiations are marked. This adaptive, creative result is called the **founder flush**.

Parapatric speciation

Variation throughout the range of a species and resulting local adaptation will create genetic gradients. Once such a cline starts, gene flow across the gradient may decrease, especially if dispersal is poor, there is selection against hybrids and increasingly pure types straying into the wrong side of the cline. The more pronounced the differences become, the more the trend is reinforced. A true hybrid zone will develop and, if reproductive isolation becomes established, eventually will disappear leaving two adjacent species. Distance over the range of the populations is the creative force, since genetic variation driven by genome mechanisms alone can create clines even across a homogeneous habitat. Geographical and ecological disruption can aid and abet progress.

Sympatric speciation

Sympatric speciation involves no spatial separation of population. Separate, cohesive genotypes are established and maintained while in contact with one another. Models of sympatric speciation have relied heavily on selection.

- **Disruptive selection** by selection of different phenotypes controlled by one gene and elimination of intermediates. The selection of phenotypes may be due to ecological interactions such as predation. Once different phenotypes are established natural selection will encourage reproductive isolation through habitat selection or positive assortative mating (different phenotypes positively choose to mate with their own kind). Sympatric speciation linked to habitat selection has been suggested as a major cause of insect diversity. (See Figure 2.11.)
- **Competitive speciation** is a variant of disruptive selection. Competition for resources within a species selects for different phenotypes able to avoid intense competition while intermediate types are lost.
- **Polyploidy** relies on the multiplication of whole sets of chromosomes so that offspring have double or more than the normal complement. Gametes of polyploids typically contain more than the normal one-half set of chromosomes so if fertilised

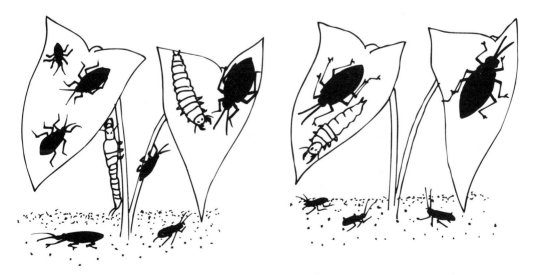

Figure 2.11 *Sympatric speciation by disruptive selection. A species of beetle, which shows wide variation in body size, shares part of its range (on the plant) with a predator. The predator can overpower small but not large beetles. Eventually large and small beetle populations exist as a result of predator selection. If differences between the beetle populations are consolidated so that they no longer interbreed successfully two new species exist where there had been one*

(whether by another polyploid or not) the offspring has a different number of chromosomes. Polyploidy relies on abnormal cell division to create the overloaded gametes. **Autopolyploidy** occurs by fertilisation between polyploids within a species. The resulting offspring often suffer low fertility, their mass of chromosome literally tangled during cell division to create gametes but vegetative reproduction can be strong. The potato is a good example. **Allopolyploidy** relies on hybridisation between different species of polyploids. Offspring are usually fully fertile, with gamete production smoothed as the chromosome contributions of each parent separate out with their own kind, avoiding tangles. Many wheat varieties are allopolyploids. Polyploids are often bigger and more productive than their forebears. Polyploidy is rare in animals but a major speciation mechanism in plants with over 43 per cent of dicotyledon and 58 per cent of monocotyledon species polyploid.

Other potential sympatric mechanisms include chromosomal rearrangements that result in reduced fertility of heterozygotes. Meantime homozygotes, potentially displaying distinctive features, can establish populations which become cohesive, much like disruptive selection. This is called **stasipatric** speciation. **Macrogenesis**, the dramatic mutation to produce a hopeful monster is a possibility, demonstrable in some laboratory populations but with little support in nature.

The relative importance of all mechanisms is still hotly debated. Examples occur of all types, their precise role dependent on the taxa in question.

The imbalance of nature

The balance of nature has a strong hold on our imagination and conveys a cosy vision of harmony. However, nature's creative imperative is anything but balanced. The diversity of life around us, at least the most familiar aspect of species richness, can be reduced to a very simple equation. Species diversity equals speciations minus extinctions. Since there are still very many species alive on Earth nature is profoundly imbalanced, with speciation generating a surplus over and above the ravages of extinction. Quite how big that surplus is remains one of the most difficult questions facing biologists. The diversity of life on Earth is still surprisingly poorly known. Chapter 3 provides an inventory of biodiversity, reviews progress and problems and tries to quantify the remaining surplus of nature's creativity.

Summary

- Ecological and evolutionary sciences seek to explain patterns and processes of biodiversity.

- Ecological factors controlling diversity range from large-scale influences of time and geography to intimate interactions between species and species' own attributes for dispersal.

- The diversification of genetic material ultimately spawns new species.

Discussion questions

1 Why do we believe in a 'balance of nature'?

2 How do wild animals in your country engineer their habitat?

3 Why is the biodiversity of tropical rainforests so great?

See also

Ecological causes of extinction, Chapter 4.
Genetics and extinction, Chapter 4.
Evolution through deep time, Chapter 1.

General further reading

The Diversity of Life. Edward O. Wilson. 1992. Harvard University Press, Harvard.
Beautiful exploration of ecology and evolution and the creation of biodiversity plus human impact.

Biological Diversity. The Coexistence of Species on Changing Landscapes. Michael A. Huston. 1994. CUP, Cambridge.

Species Diversity in Time and Space. Michael L. Rosenzweig. 1995. CUP, Cambridge.
Huston and Rosenzweig's books are very up-to-date reviews of ecological processes and how these influence biodiversity. They are also very different in emphasis, Huston concentrating on the role of disturbance and interactions, Rosenzweig on area.

The Blind Watchmaker. Richard Dawkins. 1986. Longman, Harlow.
Richard Dawkins tour de force explaining evolution's creative power.

'Generation, maintenance and loss of biodiversity'. R. Barbault and S. D. Sastrapradja. 1995. In V. H. Heywood (ed.). *Global Biodiversity Assessment*. Section 4. CUP for UNEP, Cambridge. Detailed review of ecological and evolutionary processes creating biodiversity.

3 An inventory of planet Earth

Discovering and defining the richness of life is central to our knowledge of biodiversity. This chapter covers:

- **Defining types of biodiversity**
- **Quantifying biodiversity**
- **Global patterns of biodiversity**
- **The role of ecosystems using the example of wetlands**

Biodiversity strikes a Pavlovian nerve, conjuring up images of rare mammals and rainforests but it embraces much more than these. There are three main components to biodiversity; genetic, organismal and ecological. The **genetic and subcellular diversity** includes all biodiversity expressed within individual cells plus non-cellular organisms such as viruses. The diversity of genetic information is central to this category but the variety of metabolic pathways and molecular biology of life also represent important diversity. **Taxonomic diversity** is dominated by our focus on species. Species are but one level at which organisms can be classified and the diversity of other categories such as families or phyla provides additional insights. **Ecological diversity** includes whole communities, habitats and ecosystems, including domestic stocks. Recent definitions, e.g. UNEP's Global Biodiversity Assessment, have added **cultural biodiversity** as an explicit concept, human social systems intimately dependent on the ecological system within which they exist. This has created some tensions. In 1996 Makah Indians attended the International Whaling Committee meeting to request permission to hunt five Grey Whales a year, after a 70-year gap, to help maintain their cultural system. Their request was seized upon by Norway, keen to reopen commercial whaling. Small Norwegian coastal towns that would benefit from commercial whaling were compared to the Makah Indians. **Interaction biodiversity** has also been coined to embrace

Table 3.1 *Recent categories and definitions of biodiversity*

Author	Genetic and subcellular	Taxonomic	Ecological
Eldredge (1992)	Genealogical	Phenotypic	Ecological
Groombridge (1992)	Genetic diversity	Species diversity	Ecosystem
HMSO (1994)	Genetic variation	Diversity of species	Diversity between and within ecosystems
Hawksworth and Harper (1995)	Genetic	Organismal	Ecological
Heywood (1995)	Genetic	Organismal	Ecological

the interactions between species as a fundamentally important factor in natural systems. The very success of the term biodiversity, its rise to prominence and increasing breadth of topics have fuelled some dissension, with scientists accused of pushing the term as a technological, even mythic concept, because of its allure for research funding. (See Table 3.1).

Types of biodiversity

Genetic and subcellular

Genetic, molecular and metabolic diversity are often overlooked. They represent the founding diversity of life, not just as it is today but echoes of the ancient past and potential futures. All three generate diversity within and between species and ecosystems.

Genetic diversity provides the core differences that divide life into its major types, especially important when visible structures and shape provide no reliable guide. This genetic diversity is witness to the deepest divisions of life but also some shared characteristics, stretching across huge taxonomic distances. Such genetic characteristics speak of the relatedness of all life forms one to another. Genetic diversity is the ultimate divisor and link. Genetic and metabolic diversity are especially important to understanding time, evolution and taxonomic characteristics of life. Genetic material is the raw material from which future biodiversity will be spawned.

Genetic diversity is made up of **genes**. A gene controls the expression and development of a particular feature of a living organism. The precise form of the character will vary (e.g. hair or eye colour in humans) and each of the variations of the gene is called an **allele**. So a gene in an organism will be one allele from a larger set. Genes are built of **deoxyribonucleic acid** (**DNA**), a linear molecule built like a ladder and twisted into the famous double helix. The rungs of the ladder consist of molecules called **nucleotide bases**, two joined to make each rung. There are four different bases, cytosine, guanine. thymine and adenine, which pair up to build the rungs as base pairs. Cytosine pairs with guanine, adenine with thymine. Each set of three base pairs makes up a triplet and each triplet is the code sequence for an amino acid. There are 64 permutations in which three consecutive base pairs can occur, more than enough to code for the 20 amino acids common to all organisms. So the diversity of the genetic code is a hierarchy starting with the sequence of individual rungs then triplets, the order and length of sets of triplets, the resulting alleles and quantity of genetic material in different organisms. In addition to the active genes there is extra DNA apparently doing nothing.

The result is that genetic diversity even within a single species is so vast that the information is much larger than the total number of individuals. The measurement of genetic diversity is a rapidly improving area. Box 6 outlines methods and measures.

Box 6

Genetic biodiversity: Methods and measures

Methods

- **Protein electrophoresis** Widely used since the 1960s, this technique analyses the different proteins which in turn reflect the different alleles in an individual.
- **Restriction site mapping** A recent advance relying on very specific bacterial enzymes that restrict viral damage to DNA by cutting out damage at specific points. Analysis of these cut points, called restriction sites, allows very precise analysis of gene sequences.
- **DNA and RNA sequencing** Another new approach allowing analysis of all DNA. A frequent alternative is ribonucleic acid (RNA) found in ribosomes (rRNA). Ribosomes are cell organelles that read genetic data and manufacture proteins based on these instructions. 16s rRNA (the name refers to the number of subunits in the RNA and that it is ribosomal) has been particularly important.

Measures

- **Percent polymorphic loci, P** A locus is the position of a gene on the genome. A gene where the frequency of the most common allele is <95 per cent of the total is regarded as polymorphic.
- **Number of alleles, N** This measures not only if a gene is polymorphic but also the number of alleles per gene.
- **Heterozygocity, H** Frequency of alleles, how many there are and the frequency of each, e.g. there may be three alleles, one occurring 85 per cent of the time, a second 10 per cent, a third 5 per cent.
- **Number of segregating sites, S** Restriction site positions can vary in a gene. This detail can be added to the first three items.
- **Allele tree** An attempt to construct evolutionary relationships, providing a sense of relatedness and uniqueness. In much the same way as a rare species can be picked out from among its more common kin, allele trees try to pick out the genetically unusual.

The metabolic diversity of life, particularly among bacteria, is commonly overlooked, but represents a fundamental and ancient diversification (see Box 7). The metabolic abilities of bacteria can be directly useful to humans, e.g. fermentation, or indirectly as part of wider geochemical cycles that are ecosystems services maintaining the health of the environment. Metabolic diversity can be divided between utilisation of different energy sources, energy release (respiration) and nitrogen fixation.

Taxonomic diversity

The number of species alive on Earth today is often thought of as synonymous with the term biodiversity. Work at the species level has practical advantages. Funding bodies (often the general public) recognise species. Conserving a species is a simpler concept than conserving a species' gene pool. Species projects are often easier to organise than conserving whole habitats. However, the species is but one category in a hierarchy of classification. Other categories are important measures of biodiversity, sometimes providing different contradictory messages to species level analyses.

Box 7

Microbial metabolic diversity

Tapping environmental energy sources

Energy from inorganic chemistry (chemoautotrophy)

- Sulphur bacteria. Oxidise sulphur compounds, e.g. hydrogen sulphide, to release energy used to build carbon-based food.
- Iron bacteria. Oxidise iron compounds, e.g. iron ore, to release energy used to build carbon-based food.
- Nitrifying bacteria. Oxidise ammonia to release energy used to build carbon-based food. Nitrite and nitrate byproducts are a valuable input of nitrogenous plant nutrients.
- Hydrogen bacteria. Oxidise hydrogen to release energy used to build carbon-based food.

Trap light energy (photoautotrophy or photosynthesis)

- Athiorhodacea. Split organic matter as hydrogen ion donor.
- Purple and green sulphur bacteria. Split hydrogen sulphide as hydrogen ion donor. Waste product is sulphur dioxide.
- Advanced photosynthesis. Split water as hydrogen ion donor. Waste product is oxygen.

  Note that many bacteria can trap energy by photosynthesis and by oxidation of inorganic molecules. The ability to use either method is called mixotrophy.

Metabolic energy release

Anaerobic

- Fermentation. Partial breakdown of carbon compounds.
- Anaerobic photosynthesis. Organic matter broken down as hydrogen donor for photosynthesis generates energy.
- Nitrate reducers. Split oxygen off nitrate and nitrate. Nitrogen gas produced.
- Carbonate reducers. Split oxygen off common carbonate compounds, e.g. calcium carbonate. Waste product, methane, has a possible role as a greenhouse gas.
- Sulphate reducers. Split oxygen from sulphate. Create rotten egg, hydrogen sulphide smell of fetid wetlands.
- Iron reducers. Alter chemical bonds of iron compounds resulting in energy release analogous to oxygen splitting of other anaerobic systems but without oxygen.

Aerobic

Oxidation of organic matter. Primary energy release systems for protistan, fungal, plant and animal kingdoms.

Nitrogen fixation

Bacteria extract atmospheric nitrogen gas and make it available to nitrogen cycle. Major source of nitrogen input into ecosystems.

Systematics is the branch of biological science responsible for recognising, describing, classifying and naming organisms. Part of this work involves a **taxonomic** (taxa = category, nomic = name) hierarchy, including species and other levels at which organisms can be described. Linked to this are taxonomic nomenclatures (naming systems) which provide internationally agreed rules on how species are to be named. International Codes exist for animals, plants and bacteria, plus a provisional system for viruses. The taxonomic classifications also portray our understanding of the evolution and relatedness of life, the **phylogenetic classification**. Humans have a natural tendency to classify things, biodiversity being no exception. Classification depends upon what we can observe and what we think it means (often influenced by social, political and religious attitudes). Taxonomic classifications of life have varied throughout time. Box 8 outlines the development of schemes in Europe.

Box 8

Classifications of life through European history

Classifications of life have changed over time, but even ancient systems can show precise methodologies and criteria. Changes reflect use of increasingly small-scale, internal features (internal organs, internal cell organelles, internal organelle RNA).

Greek Aristotle, in *Historia Animalium*, 486 BC, classified animals by clearly stated criteria for grouping animals together; (1) 'With regard to animals there are those which have all their parts identical . . . specifically identical in form'; (2) 'When other parts are the same but differ from one another by more or less, then they belong to animals of the same genos. By genos I mean for example bird or fish'; (3) 'There also exist animals whose parts are neither the same by form but by analogy'.

Linneaus The famed Swedish biologist tried a global classification of animals using multiple criteria. Life forms must be distinguished by external criteria, e.g. skin, locomotion. Each life form should have roughly equal roles in the economy of nature. Within each life form criteria used for subgroups should be essential for finding or processing food, e.g. teeth for mammals, beak for birds. Linneaus' work resulted in classifications of plants based primarily on reproduction mechanisms and of animals based on feeding.

Cuvier Through the late eighteenth and early nineteenth century Cuvier pioneered comparative anatomy, especially the internal resemblance of animals, as a means of classification. Cuvier saw some characteristics as fundamentally more definitive than others, notably the nervous systems, though practicalities favoured use of bones. Cuvier broke the vision of animal life as a ladder of progression, splitting animals into four lineages (vertebrates, molluscs, articulates, e.g. insects, and radiates) based on form and function. This split is still echoed in current classification of animal lineages.

Whittaker's five kingdoms By the mid-twentieth century advances in cell biology led to a revision of life into five main kingdoms, defined primarily by subcellular features, e.g. cell walls, links between cells, organelles and mechanisms of cell division. The five kingdom system (Monera, Protista, Fungi, Plantae, Animalia) is still widely used. (See Figure 3.1.)

16R rRNA By 1990 analyses of DNA and RNA, particularly work on 16s rRNA sequences suggested a fundamental revision of the five kingdoms. Shared and different 16s rRNA sequences suggested life divided into two main branches, the Bacteria on one hand and a second branch itself split between the Archaea (previously thought of as close allies of the Bacteria) and the Eucarya (everything else).

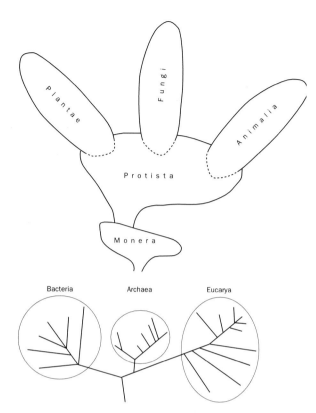

Figure 3.1 *Whittaker's five kingdoms of life and recent three domain revision based on RNA. The Bacteria and Archaea domains were both part of Whittaker's Monera, while the new Eucarya domain encompasses what were the Protista, Fungi, Animalia and Plantae Kingdoms*

Source: Redrawn from Heywood 1995.

Recently systematics and taxonomy have not been popular fields with biologists or funding bodies but burgeoning biodiversity studies have focused attention on our woefully incomplete catalogue of life. 'You don't know what you've got 'til it's gone', the 1960s environmental protest song lamented. We now realise that you don't even know then. An urgent task in biodiversity research is to rebuild skills in systematics. Table 3.2 gives the typical classification hierarchy, from species up to the fundamental divisions of life into kingdoms or domains. Note that there is marked variation in diversity even at very high levels of classification. Recognised numbers of phlya in the five kingdom classification of life are: Prokaryte 16; Protista 27 (at least); Fungi 5; Animalia 32; and Plantae 9.

Defining species

The **species** remains the focus of most biodiversity research but there is no precise simple universal definition of what a species is. Some alleged species fit all criteria of all definitions, others are hard to fit to any.

The word species literally means outward or visible form. Classifications up to the twentieth century relied on the physical, often outward, similarity of features to distinguish a species. The use of visible characters to define a species is the **morphological species concept**. This approach is widely used with collections of dead specimens or fossils which can provide no other information on breeding and ecology. Species with varied individuals, e.g. camouflaged, mimics or variable growth forms, risk being split into several species.

Insights from evolution and genetics prompted the **biological** (or **isolation**) **species concept**. A species consists of populations of interbreeding individuals, able to reproduce successfully with other populations. Central to this definition is the idea of reproductive isolation, be it a physical barrier or behavioural, physiological or genetic inability to mate and produce fertile young. The isolating mechanisms can be pre- or post-mating. Pre-mating includes separate habitat, time, behavioural and physiological mismatch, mechanical inability to mate and gamete inability to

Table 3.2 *The taxonomic hierarchy. Each level in the hierarchy is a category at which fundamental features are shared. Each category will consist of one or more of the next category down (e.g. a genus consists of one or more species). Sometimes subcategories are used (e.g. subclass, suborder). Recent analysis of genetic biodiversity has caused a rethink of what had seemed the ultimate, kingdom-level divides, a salutary lesson*

Taxonomic hierarchy	Human beings	Mount Kupe Bush Shrike (Chapter 4)	African Elephant (Chapter 4)
Kingdom or Domain	Kingdom Animalia. Domain Eucarya	Kingdom Animalia. Domain Eucarya	Kingdom Animalia. Domain Eucarya
Phylum or Division	Chordata	Chordata	Chordata
Class	Mammalia	Aves	Mammalia
Order	Primates	Passeriformes	Probosidea
Family	Hominidae	Corvidae	Elephantidae
Genus	*Homo*	*Telophorus*	*Loxodonta*
Species	*sapiens*	*Telophorus kupeensis*	*africana*

fertilise. Post-mating includes hybrid inviability (embryos or young die), hybrid sterility (young are born but cannot reproduce) and hybrid breakdown (young mature and breed but their offspring are less viable). This definition does not work with asexual species. A surprising number of recognised species living side by side appear to mate and exchange genetic material without the species involved coalescing into an indistinguishable mélange. Many plants show interbreeding. Tests of the American Buffalo genome have revealed domestic cattle genes but no morphological blur. Species defined by breeding are also difficult for systematists. Not only are the breeding abilities of fossils impossible to fathom but over evolutionary time species change, so much so that they might be morphologically distinct but at every step interbreeding between immediate parent–offspring generations was possible. The morphologically distinct ancestor and descendent species are linked by a seamless reproductive thread.

Refinements to the biological species concept include: the **recognition species concept**, emphasising not the barriers to reproduction but shared characteristics permitting reproduction; the **cohesion concept**, a species defined by cohesive mechanisms that prevent interbreeding with other species and evolutionary diversification within the species; and the **ecological concept**, species defined by occupation of a niche unique compared to other species in its range and evolving separately from populations beyond its range. All remain flawed by hybridisation and problems of practical measurement.

A third approach is the **evolutionary species concept**. This emphasises the **phylogenetic concept** with species defined as the smallest population (sexual) or lineage (asexual) diagnostically distinct from other such populations and with a discrete lineage. The diagnostic characters are often genetic. When applied to species recognised by morphological or biological concept criteria, the sheer genetic diversity within species often suggests they should be splintered into many separate lineages. The evolutionary concept defines a species by a single distinct lineage with an identifiable evolutionary history and fate.

Ecological diversity

Taxonomic diversity classifies types of organisms and their relatedness but organisms do not live in isolation from one another or the physical world. Humans have long recognised different ecosystems of apparently interdependent life. Ecological science focuses on these patterns and processes, hence ecological diversity as the inclusive term for this third category.

Ecological diversity covers a host of concepts: ecosystems, communities, assemblages, habitats, biomes and biogeographical regions. These are not one and the same, indeed some may not be biodiversity proper. The term **ecosystem** embraces the living organisms and non-living (abiotic) features such as climate and geology of a site. Some have argued that the inclusion of abiotic components excludes ecosystems from biodiversity. A **community** refers to organisms living together, essentially the live component of an ecosystem. Again this deceptively simple idea is problematic. The term implies a linkage, an interdependence of species that may not exist. Even tightly linked communities will harbour fleeting tourist species, moving through without necessarily playing any role while other communities may be very loosely tied assortments, often described as an **assemblage**. Habitat conjures up precise images, e.g. Giant Panda habitat, but the term may not mean anything if the species is not present. Does Giant Panda habitat cease to exist should the Giant Panda become extinct? **Biome** is the term associated with global or continental scale, regional ecosystems defined by vegetation and fauna, in its turn largely determined by climate. A **formation** is a similar concept, relying solely on vegetation data. Even colloquially familiar biome terms such as rainforest or wetland become difficult to define precisely due to global variations or to give sharp boundaries where they grade into each other.

Ecological biodiversity also includes geographical foci such as high diversity hot spots, continental and oceanic islands, plus regions of endemism. If ecological diversity (whether ecosystem, community or any other tricky concept) was nothing more than the sum of its parts (e.g. the species list) this category of biodiversity could be largely ignored, but ecological diversity has its own importance. Ecological patterns emerge at the ecosystem and community level which are more than the sum of the parts of the species present. Ecosystems can (albeit rather mechanically) be said to provide ecosystem services (or functions) such as nutrient and gas cycling. An ecosystem that loses a species may not be merely minus one species but may function in a very different way with knock-on consequences to other ecosystems and perhaps globally. So the ecological category of diversity is important.

Classifications of ecological diversity vary with scale, just like the genetic (nucleotide molecules up to differences between populations) and taxonomic (subspecies to domains). Three main scales are commonly used for ecosystems:

global, regional and national. Global classifications rely on major vegetation types (biomes) typically defined by combinations of dominant vegetation, landscape and climate, sometimes plus biogeographical position. Regional classifications are often based on practical definitions. National classifications work at a finer scale, often based on precise species presence and abundance. Classifications of wetlands global, regional or national, are good examples of ecosystem categorisation, the difficulties of definition and how resulting classifications can vary with the purpose of the systems used.

Wetland classification

Global classifications

At the gross global level wetland habitats are subsumed within the broader biomes. A widely cited definition of wetlands is from the RAMSAR convention (the treaty is outlined in Chapter 5):

> Areas of marsh, fen, peatland or water, whether natural or artificial, permanent or temporary with water that is static, flowing, fresh, brackish or slat, including areas of marine water the depth of which at low tide does not exceed six metres.

Wetlands are characterised by the presence of water at the surface or in the root zone, unique soil conditions and hydrophyte (water loving) plants. Distinguishing different types of wetlands is problematic. The duration and depth of inundation vary. Wetlands are often marginal habitats between truly aquatic and terrestrial systems. Size can vary from small ponds to huge regional expanses. Definitions also vary according to the aim of the classification scheme and cultural traditions, e.g. in the USA swamp means forested, in Europe typically not so. At a global scale nine main wetland types are commonly distinguished using general biome and landscape characters.

1 **Bogs**. Acidic, peat-building wetlands with water and nutrient input from precipitation. Typically in wet, cool climates and characterised by *Sphagnum* mosses.

2 **Fens**. Peat-building but with nutrients and hydrology influenced by surrounding landscape and geology, neutralising acidity. (Note that bogs and fens are often combined together as **mires**, a generic term for peat-based wetlands).

3 **Swamps**. So waterlogged that water usually covers surface and dominated by a few species, e.g. reeds or characteristic trees, e.g. swamp cypress.

4 **Marsh**. Inundation often seasonal. Diverse herbaceous vegetation with little or no peat accumulation.

5 **Floodplains**. Periodic overspill from lakes and rivers. Very varied.

6 **Shallow lakes**. Open water up to a few metres deep.

7 **Salt marsh**. Tidal herbaceous coastal sward of temperate latitudes.

8 **Mangroves**. Tidal coastal woodlands between 32°N and 30°S, characterised by mangrove trees. Exclude saltmarsh from tropics by competition but limited to tropics by climate.

9 **Anthropogenic**. Wetlands created by humans, e.g. paddies.

National classifications of wetlands

At a national level much more detailed classifications can be devised, often differing depending on the purpose for which they are used. In the USA there are over 50 wetland classification systems in use for water regulation, recreation and wildlife conservation. Here is an example from the UK based on detailed botanical classification.

Classification by species and floristic composition: the National Vegetation Classification (NVC)

The NVC is a systematic and comprehensive classification of UK plant communities. Vegetation types have been distinguished by computer analysis of the presence and absence of individual species and their relative abundance. The classifications for four habitat types (woodland, mires, grassland and aquatic) are available. All four contain communities identified as wetland, 100 in all with coastal ecosystems still to be added. Each community type is described in detail, listing plant species, abundance, physical and chemical conditions in the habitat and distribution. Each type is given its own code and name, derived from the diagnostic vegetation, e.g. S9 *Carex rostrata* swamp is the ninth recognised swamp community from the aquatic habitats, characterised by the bottle sedge, *Carex rostrata*.

- Woodland. Seven wet woodland communities.
- Mires (bogs, wet heaths and fens). 38 wetland communities.
- Grassland. Three inundation communities.
- Aquatic. 24 submerged and 28 swamp/tall herb fen communities.

Regional classification of wetlands

The SADCC Wetlands Programme is a wetlands inventory for southern Africa. In 1991 the SADCC Wetlands Conservation Conference for southern Africa, meeting in Gabarone, capitol of Botswana, set up a programme intended to produce a comprehensive inventory for the ten countries of Angola, Botswana, Lesotho, Malawi, Mozambique, Namibia, South Africa, Swaziland, Zambia and Zimbabwe. The driving force was increasing recognition of the value of wetland ecosystem biodiversity. Wetlands have the highest biological productivity compared to equal areas of other ecosystems in the region. They offered multiple uses and harboured the greatest taxonomic biodiversity. The programme collects baseline data on wetland area, distribution seasonality, characteristics and value at a national level. The national scale is important to draw up reasonably clear boundaries and to detect small and unusual combination wetlands.

Sources of data included large-scale satellite remote sensing, aerial photographs, maps and vegetation surveys, as well as the collation of existing reports on individual wetlands. Highly detailed botanical surveys are lacking or insufficient to create classifications as detailed as the UK's National Vegetation Classification but this is not a serious problem given the purpose of the SADCC inventory. The inventory work also produced practical benefits, improving co-operation across political borders that cut through wetlands and between different interest groups, e.g. farmers, foresters and conservationists. The final aim is threefold: to quantify the extent, types and uses of wetlands; to quantify the rate and extent of alteration and loss; and to disseminate the information to managers and users of these precious ecosystems. Wetland classifications in Zimbabwe reflect the different ecosystems and human concerns.

Zimbabwe's wetlands ocupy about 3.3 per cent of the country. No systematic classification exists as yet but the broad categories are: swamps, permanently inundated; floodplains and flats, seasonally inundated; dams, manmade, small impoundments important for irrigation; pans, seasonal waterholes sometimes supplied by pumped water for cattle or wildlife tourism; dambos, small wetlands widely used for horticulture and lakes. Zimbabwe is landlocked and has no coastal ecosystems.

Domestic biodiversity

Biodiversity is widely thought of as an entirely natural phenomenon but definitions now recognise the importance of domesticated species, ecosystems forged from human activity and some cultures intimately tied to their environment. Biodiversity plays a role in the ethical, religious and social values of societies. Differences between cultures affect our valuation of biodiversity, in turn affecting conservation. The fate of biodiversity in different cultures may provide important insights since humans are now the dominant influence on remaining species and ecosystems. Several indigenous cultures are (or were) founded on an ecological intimacy with biodiversity. The native plains Indians of North America relied on the buffalo herds, not only as food but as an icon central to their beliefs and customs. The Marsh Arabs of Iraq have a social and agricultural system dependent on their wetland home.

Domestic biodiversity is a tiny fraction of the whole but provides 90 per cent of human food supplies. Domestic stocks include life from all kingdoms. In addition landscapes created by humans, whether farmland, forests, gardens and urban landscapes, are new ecosystems. The extent of human interference with domestic biodiversity varies. Truly domestic species have been subject to marked artificial selection, so that many characteristics will be markedly different to wild ancestors, e.g. cattle, sheep and pigs. Domesticated species (exploited captives) retain many similarities to wild brethren, e.g. reindeer, yak, camels. Feral populations are domesticates that have reverted to living wild. They can create problems, e.g. feral cat predation of native marsupials in Australia, but are unique additions to wildlife.

Domestication leads to a proliferation of forms, **cultivars** in plants and **breeds** in animals, which can be defined by heritable, distinct and uniform characteristics different to other such groups. Plant domestication spans five categories, first the wild ancestor stock. Weedy relatives often flourish in marginal habitats impacted by humans, a genetic bridge between wild relatives and the truly domestic. Primitive cultivars are genetically diverse stock still open to natural selection. Modern cultivars are the result of purely artificial selection of supermarket qualities. Advanced breeding lines are the recent additions, their genomes created by laboratory manipulation, perhaps including genetic diversity impossible through natural processes.

Numbers of domesticated species are small. Of the 320,000 vascular plants, 500 are domesticated and 15–20 are major crop species. The 50,000 vertebrates (excluding fish) have provided 30–40 domesticates, plus 60 semi-domesticated. Another 200 fish and invertebrates are commonly used in aquaculture and four insects (honey bee, various silkworm caterpillars and the cochineal bug). Fungi and microbes are grown commercially for food, brewing and biotechnological uses, e.g. pharmaceuticals. The lack of species is made up for by breeds. Estimates of total numbers of breeds vary, e.g. cattle 780–1090, sheep 860–1200, pigs 260–480 (if we cannot keep track of domestic diversity is it any wonder that numbers of wild species are so difficult to estimate). The core domestic farm mammals (cow, buffalo, horse, ass, pig and sheep) comprise 3,237 breeds of which 474 are rare and an additional 617 have gone extinct since the end of the nineteenth century.

The variety of breeds is a resource. Many have adaptations to local conditions. These adaptations can be exported to new ranges either by moving the animals or controlled breeding. The known traits of breeds allows easier control of artificial selection and discovery of new mutations. The variety is also insurance against the unforeseen and is being eroded by the globalisation and intensification of agriculture. Losses are biased towards the developed world which may have nurtured more breeds and recorded extinctions. Table 3.3 gives global, regional (Africa and Europe) and national (Zimbabwe and UK) cattle and sheep breed diversity and losses.

Table 3.3 *Surviving and extinct cattle and sheep breeds. Numbers given for global, European, African, UK and Zimbabwean records*

	Global	Europe	Africa	UK	Zimbabwe
Cattle					
Extant total	732	209	168	42	4
Extant rare	141	154	10	16	0
Lost	224	101	22	5	3
Sheep					
Extant total	917	356	131	73	3
Extant rare	135	109	1	17	0
Lost	146	97	4	8	0

Source: Groombridge 1992 and Heywood 1995.

The inventory of domestic diversity also includes more ominous items. Manipulations of genes in some animals and plants have created hybrids which would probably be impossible through natural selection. Incorporation of genes from one species into another is now commonplace, creating **transgenic species**. Transgenic sheep, pigs, cattle, horses and rodents have been created containing genes to enhance obvious features, e.g. mouse genes controlling fur structure in sheep enhance fleece productivity. Human genes have also been incorporated into some animal species to produce medical products, e.g. growth hormones. Genes can even be combined between kingdoms with bacterial genes incorporated into some mammals. Such advances now include the ability to coalesce species, most famously the recent goat/sheep hybrid, an example of a **chimera species**.

Disease eradication programmes have also created a dilemma for conservation. One virus, smallpox, once a deadly killer was declared extinct in the wild by 1980. Two stocks remain in laboratories, one in the USA one in Russia. In 1995 the World Health Organisation (WHO) voted not to destroy these stocks. Had the destruction been authorised this would have been the first ever intentional extinction of a species. In 1996 the WHO recommended destruction in June 1999 following a three-year search to check for forgotten or hidden stocks, nervous of illegal stockpiles kept for biological warfare. Instead the intrinsic value of the virus as well as possible future need for research in the event of the rediscovery of a wild stock or near relative brought a reprieve.

Chimeras and deadly viruses are biodiversity's stranger corner but our attitudes to nature have always been coloured by the culture of the day. One Anglo-Saxon bishop issued a letter to his parishes scolding the populace over their fear of werewolves which were, after all, a common feature of the countryside.

Measuring biodiversity

Quantifying the variety and richness of life on Earth is central to studies of biodiversity. This following section provides estimates of taxonomic and ecosystem diversity. But be warned. The numerical data are our best guesses, not definitive answers. For example, Table 3.3 of surviving cattle and sheep breeds disguises serious differences between estimates. If we cannot be sure about numbers of animals that we control, estimating numbers of wild species is even more difficult.

The numbers of known species

Inventories of the diversity of species on Earth today can use known species and estimates of likely numbers (including likely maxima and minima). The numbers known and reliability of estimates vary greatly with taxa. Birds and mammals are well documented, though there are still some surprises, e.g. the Udzungwa partridge, *Xenoperdix udzungwensis*, a bird species first identified in 1991, in a dinner (Box 9).

Numbers of many invertebrates and micro-organisms remain darkly mysterious. Estimates of the rate at which new species are discovered can provide some insight into these voids. Table 3.4 gives estimates of known species, showing how these have changed through time.

Box 9

A new species of bird, the Udzungwa partridge

Our knowledge of bird species is unusually complete but new species can be found.
In 1991 two Danish zoologists were exploring Udzunga Mountain, part of the endemic rich
Eastern Arc Mountains of Tanzania. One night at supper the expedition cooks produced
'chicken stew'. At the bottom of the pot were the two legs of a bird, not exactly a chicken
(African chicken stew is often not exactly chicken) but which the zoologists thought must be
one of the chickenlike Francolins. Keeping an eye out for Francolins, not previously known
in the area, subsequent sightings of the bird did not match anything in the published
literature.

During a return visit in search of the mystery bird local guides snared a male and a female.
Comparison of the specimens to museum skin collections revealed that the bird was a new
species, related to species from South East Asia. The Udzungwa forest partridge was not
merely a new species of bird, but a relict species from 15 million years ago when forests
spanned Africa to Asia. The Udzungwa partridge is a classic relict species, contributing to the
diversity of this hot spot. (See Figure 3.2.)

Figure 3.2 *The Udzungwa partridge*

Source: Reproduced with kind permission from Martin Woodcock and the African Bird Club.

Estimates of global diversity have changed markedly throughout history. It is
important to separate these counts of known taxa from estimates of the actual
numbers, which use various assumptions to include projections of likely numbers as
well as the as yet undescribed.

Table 3.4 *Numbers of described species and estimates of actual numbers for selected taxa, in thousands*

Taxa	Species described	Estimated numbers of species: high	Estimated numbers of species: low	Working figure
Viruses	4	1,000	50	400
Bacteria	4	3,000	50	1,000
Fungi	72	2,700	200	1,500
Protozoa and algae	80	1,200	210	600
Plants	270	500	300	320
Nematodes	25	1,000	100	400
Insects	950	100,000	2,000	8,000
Molluscs	70	200	100	200
Chordates	45	55	50	500

Source: Groombridge 1992; Heywood 1995.

The total described inventory for some taxa such as multicellular animals and green plants gives good minimum estimates, but there are problems even with these organisms we have described. For some there is a debate over the numbers of genuine species versus subspecies. In more diverse, typically less well-known taxa, including some of the numerically and ecologically dominant groups such as worms and molluscs (Phyla Annelida and Mollusca respectively), there is confusion because several nominal species (anything that has been described and had a scientific name attached to it) are actually specimens of the same species. Each of the names is merely a synonym for the same thing, perhaps arising because of variable shapes, names given to different stages in the life cycle, earlier names remaining unknown to later taxonomists or confusion due to reclassification. The numbers of known mollusc species varies between 45,000–150,000, with 70,000 a compromise. Extensive tangled synonyms need painstaking revision. Even without these problems the literature of described species is confused and scattered. The apparently simple task of totting up published records gives different totals in different studies. (See Figure 3.3.)

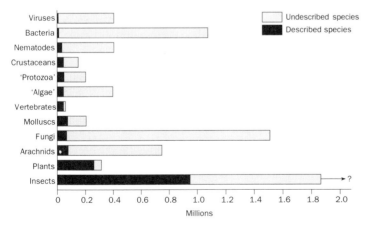

Figure 3.3 *Estimates of described and likely totals of species for selected taxa*

Source: Redrawn from Heywood 1995.

Species are still being described, even for the best known groups, and description rates provide some sense of the task ahead. Analysis of standard published records of new species between 1979 and 1988 shows remarkable consistency with between 10,912 to 11,599 new species of animal added each year. As a general figure 13,000 new species are described per annum.

Recent estimates range between 1.4 and 1.7 million species. The *Global Biodiversity review* (Groombridge 1992) settles at a figure of 1.7 million described species. UNEP's *Global Biodiversity Assessment* (Heywood 1995) states 1.75 million. But the species we know of are vastly outnumbered by those we do not. Those we do know are a biased sample, typically large attractive taxa (mammals, butterflies), those that are easy to find, in thoroughly studied regions plus pests and parasites associated with any of the above. (See Figure 3.4.) Estimating this likely total is as important and as fraught as totalling up numbers of described species.

Estimating the actual numbers of species

If we cannot agree on the numbers of species we claim to know, the chances of estimating numbers of those we do not know seems rash. However, the actual total of species on Earth today has become something of a Holy Grail in recent years. Many approaches have been used to estimate the actual total of species. Most involve using some data from a taxon, region or ecosystem and scaling up the patterns to global proportions. Each has strengths and weaknesses and usually different answers. The *Global Biodiversity* review (Groombridge 1992) settles for 12.5 million species. (See Figure 3.5.)

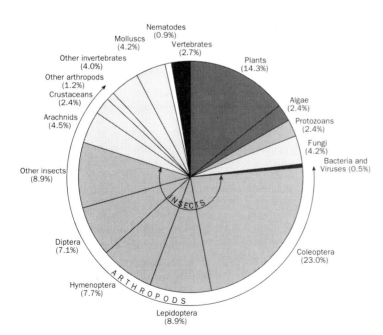

Time series

This technique uses the rates at which new species are being described and extrapolates the trends to look for an asymptote. While we may be reaching an asymptote for a few taxa, e.g. birds and mammals, there have been recent bursts of finds for many invertebrate groups so that time series seem very unreliable. (See Figure 3.6.)

Expert opinion

Specialists with intimate knowledge of particular taxa can be asked to provide their best guess of likely totals and

Figure 3.4 *Proportions of different taxa from described species. The diagram includes taxa likely to contain 100,000+ species plus vertebrates*

Source: Redrawn from Groombridge 1992.

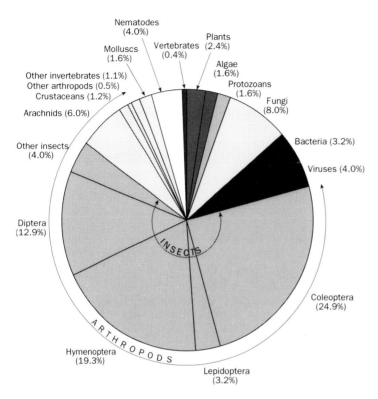

Figure 3.5 *Proportions of different taxa from conservative estimates of actual total numbers of species. The diagram includes taxa likely to contain 100,000+ species plus vertebrates*

Source: Redrawn from Groombridge 1992.

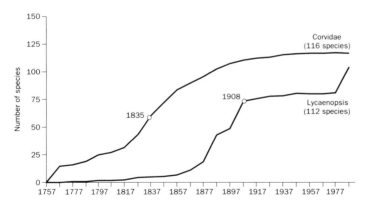

Figure 3.6 *Numbers of described* **Corvidae** *(crows) and* **Lycaenopsis** *(part of the blue butterfly family). The dates highlighted indicate points at which half the species known in 1987 were recorded. Note the limited recent additions to Crow list but bursts of additions, some recent, to the butterflies*

Source: Redrawn from Groombridge 1992.

the results summed. Expert opinion has varied greatly over time with estimates being revised ever upward. In the seventeenth century John Ray thought there were betwen 10,000 to 20,000 species while Hutchinson (1959) put the figure at one million.

Empirical relationships

There is a host of ecological patterns, often described in detail for individual communities, that can be scaled up from the original local data to global proportions (Box 10). These include the ratios of species number:body size; species:area; and species:population abundance; plus patterns derived from food webs and plant–herbivore associations. (See Figure 3.7.)

Taxon to taxon, region to region ratios

Such techniques rely on possession of good data for the numbers of taxa (both well-known and more obscure groups) in a particular region. The ratio of species of a well-known taxa (e.g. butterflies) to total taxa (e.g. all other insects) for the reliably surveyed region can then be used to estimate total numbers by

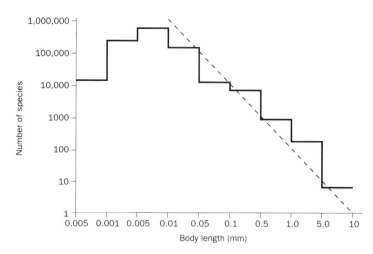

Figure 3.7 *May's 'crude estimate' of the abundance of terrestrial animals of different body sizes. The dashed line is a similar model for the relationship based on the work of Hutchinson and MacArthur. Patterns such as these can be used as the basis for estimating total species diversity, extrapolating from what is known to include poorly described taxa*

Source: Redrawn from May (1986) *Ecology*, 67.

counting the numbers of the same well-known taxa in comparatively poorly surveyed regions, assuming the same ratio of well-known taxon species to all other species holds true and multiplying up to get an estimate of the total. A classic example has been the use of butterfly to insect ratios from the UK. The UK's insect fauna is unusually well recorded, so the data are reliable. There are 67 species of extant butterfly to about 22,000 insect species in total, a ratio of 1:328.4. Globally there are some 17,500 species of butterflies. Assuming the UK butterfly:total insect ratio holds globally this gives an estimate for insect species numbers of 5.75 million (see Box 10).

Box 10

Erwin's famous 30 million species

In 1982 entomologist Terry Erwin published a famed minimum estimate of species numbers, an astonishing 30 million. Erwin's estimate was based on a study of the insect fauna found on *Luehea seemannii*, a canopy tree, in Panama. He used insecticide fog sprayed into the tree to knock down insects. He caught some 1,200 species of beetle, of which 162 were specific to the *L. seemannii* tree. He then used a series of assumptions to reach a global estimate of tropical forest insect species richness. First, that there are 50,000 tropical forest tree species and that each would harbour 162 specialist beetle species, i.e. 8,000,000 beetles. Next, that canopy beetles represented 40 per cent of canopy arthropods so total canopy arthropods = 20,000,000. Finally, that there were twice as many canopy species as ground species, so add another 1,000,000 to reach the final 30,000,000. Erwin's assumptions are rightfully open to criticism but the very magnitude of the estimate was enough to prompt resurgent interest in the inventory of life on Earth.

Problems with estimation techniques

● Scaling up   Many of the methods used rely on taking detailed data from a local, thoroughly documented site and scaling up to global proportions. Very many ecological patterns and processes are scale dependent. The patterns shown in local studies might not scale up globally in a reliable way. Patterns and processes could be diminished or exaggerated.

● Comparing taxa   These gross estimates commonly lump taxa together. Yet this assumes patterns and processes that work well for one taxa may not fit another. Data for ants might not be reliably extrapolated to estimate beetles.

● The fraction of undescribed species   Estimates rely on what we already know. Data used as the basis of a model may still be missing many undescribed species. This is a catch 22 situation: to estimate numbers of undescribed species we have to know about the undescribed numbers in the source data we use. (See Box 11.)

● Species definitions   The criteria used to define a good species vary with taxa. The minute separations used to distinguish sub and sibling species among the birds are not applied to the desperately underworked micro-organisms. These problems afflict habitats differently. Sub and sibling species separations are poorly drawn in marine studies compared to land.

● The workforce   The size, expertise and where in the world the workforce is based all affect estimates.

Box 11

Soil fauna: The other last biotic frontier

Deep sea species diversity has been cited as the last frontier that is largely uncharted. Recent data of soil microarthropods from the soils of a typically harsh but accessible habitat of sand dunes, suggest that soil fauna deserve an equal billing. Samples yielded densities of up to 1,400,000 individuals m^2, compared to previous studies of soil fauna from as low as 3,000 m^2 in Indonesian rainforest, through 110,000 m^2 from Zaïre rainforest litter to 386,000 m^2 from Alpine grassland. The numbers of species was small, only 31, but mostly undescribed. More importantly was the small size of many of the new species, less than 0.2 mm long. Many classic estimates of total species numbers rely on extrapolating known patterns of body size relative to abundance from local studies up to a global scale. The abundance of small-bodied species from this one study of dunes would be enough to increase Robert May's classic 1988 estimate of 10 million terrestrial animal species, based on body sizes down to 0.2 mm, up to 20 million.

The major gaps

The problems of bias among described taxa, estimation techniques and expertise have left major gaps in our knowledge of biodiversity, including particular taxa, habitats and conceptual topics, super diverse groups and reference sites.

Ecosystems

Marine systems
Although marine systems cover the majority of the planet our knowledge is very patchy. While the biodiversity associated with coral reefs has been compared to that of tropical rainforests, other habitats such as the surface waters of open oceans resemble deserts. Recent sampling of deep sea sediments coupled with greater interest in micro-organisms has provided evidence of very high diversity.

Tropical rainforests
Although rainforests appear to support the greatest biodiversity on Earth, research effort is still overshadowed by the sheer size of the task. Particular habitats within the forest have yielded tantalisingly diverse communities, but whether such results can be extrapolated reliably remains unknown.

Taxa

Parasites
Given that most organisms seem to suffer parasitism, it is tempting to imagine parasite diversity as vast as the range of potential hosts. This is especially so in the case of those insects that lay their eggs in or on living hosts (typically other arthropods), the host being killed only once the parasite matures and hatches. Such insects are specifically termed parasitoids, this distinction from other parasites reflecting the almost invariably fatal impact on the host, unlike other parasites. Given that insects make up the overwhelming majority of species, if every species had its own parasitoid (and parasitoids are in turn attacked by their own super-parasitoids), this automatically increases total species massively. However, evidence of parasite and parasitoid biodiversity is sketchy.

Fungi and micro-organisms
Fungi and micro-organisms are the two main under-researched groups. Difficulties of sampling, identification and concepts of species plague work with these two groups. Improving analysis of genetic characters is providing new insights.

Nematodes
Nematodes are a phylum of worms, capitalising on a very uniform body plan but extremely abundant in very many habitats. Problems from lack of expertise and identification have hindered investigation of this phylum but recent evidence hints at high species diversity.

Mites
These are members of the Phylum Arthropoda. Like the nematodes, problems of lack of expertise and identification have held back work but there are suggestions of high species diversity.

Insects

Despite the considerable work devoted to the class Insecta there remain very poorly known groups and habitats.

Conceptual gaps

Super diverse groups

Some taxa have been described as super (hyper, or ultra) diverse (or speciose), reflecting an inordinate variety of species. This accolade suggests that the patterns and underlying processes may be unusually magnified or even different to the typical determinants of diversity. Understanding such differences is important but also difficult and may require separate models to estimate total numbers.

Special species and areas

All diversity is not equal. An individual species, perhaps just one comprising an unusual lineage, might be rated as special, more worthy of protection than another with many related species in a diverse genus. How to measure the phylogenetic isolation and uniqueness of a species is a new field. Some taxa may be very useful indicators of the overall diversity of a site. Their diversity may accurately reflect the variety of other species which cannot be measured for want of time and techniques. Similarly focal taxa are groups combining concepts of the special and representative and could be useful signals of important habitats.

Essential areas and reference sites are also concepts only recently developed. The idea of **Centres of Diversity** has been developed, reflecting the evolutionary heartland (or perhaps last refuge) for a taxon. Linked to this, **complementarity** has been developed as one criterion to assess regional diversity, especially for plants. All the species of a taxon make up the total complement. The single most important site (or country) for this taxon is that with the highest proportion of this complement, the site with the highest proportion of the remaining species, the next most valuable. Conversely Centres of Diversity for birds have concentrated on aggregations of endemic species. Regions of high diversity or endemicity for one taxon are not necessarily so for another. All these ideas, their usefulness and identification are comparatively new, debatable and poorly researched.

Global biodiversity patterns

Taxonomic biodiversity

Global patterns of species diversity are well known for only a few groups. There are marked patterns, in particular concentrations of species within biodiversity hot spots, biogeographical variations between continents and trends such as the increase of species in most taxa towards the tropics. Birds and plants have been studied in detail, reflecting a bias in expertise and available data, but already producing useful insights for conservation planning.

Bird species number nearly 9,700, the most speciose of terrestrial vertebrates compared to over 4,000 amphibians, 6,550 reptiles and 4,327 mammals. Birds are one of the most thoroughly recorded taxa. Their distribution is unevenly spread across the main biogeographic realms, with 3,083 species in South America, 2,280 in Asia, 1,900 Africa, 950 in the Palaearctic, 900 in Australasia and 800 in the Nearctic. Within continents the uneven pattern continues. Table 3.5 outlines species, family and endemic diversity from the top five South American countries plus Galapagos islands. (See Figure 3.8.)

Birds have been used as a focus to identify priority areas for conservation, primarily using endemics. Birds are potential good indicators of biodiversity. They are found throughout the world and have diversified in all terrestrial regions and habitats. Their taxonomy and species distributions are well known. There is also evidence for correlation between their diversity and that of other terrestrial vertebrates and vascular plants. The International Council for Bird Preservation has run an extensive global programme to identify Endemic Bird Areas (EBAs). (See Plate 12.) The survey relies on endemics, or more exactly restricted range species breeding over a total range of no more than 50,000 sq. km. Any areas with two or more restricted range species are cited as EBAs. Of birds 2,609 fitted the range criterion, plus 59 known to have gone extinct since 1800. The analyses suggested 221 EBAs, 70 per cent in the tropics, embracing 2,480 species, 77 per cent of known threatened bird species. EBAs are ranked into three categories: Priority I (most important/threatened) II and III (less important/ threatened). These categories include estimation of biological importance (restricted range species relative to area, taxonomic uniqueness and known significance to other plant and animal taxa) and degree of threat (threat to birds, proportion of area within IUCN recognised protected areas). Table 3.6 gives numbers of EBAs, bird species and numbers in priority categories for the five countries containing the largest number of EBAs. (See Figure 3.9.)

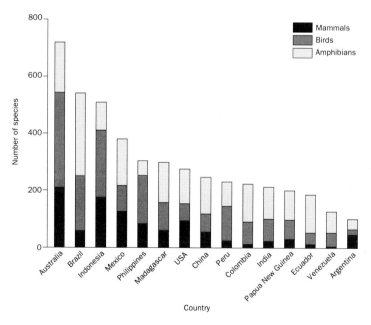

Figure 3.8 *Top fifteen countries for endemic mammals, birds and amphibia*

Source: Data from Groombridge 1992.

Table 3.5 *Bird taxonomic diversity of top five South American countries plus Galapagos islands. Note Ecuador's high diversity despite small size and Galapagos' proportion of endemic species*

Country	Area, km^2	Species total	Endemic species	Family total	Endemic families	Vascular plants, species numbers
Colombia	1,138,914	1,700+	59	79	20	35,000
Peru	1,285,216	1,700	104	86	23	13,000
Brazil	8,511,965	1,661	179	84	23	55,000
Ecuador	283,561	1,550	12	82	21	16,500–20,000
Venezuela	1,098,581	1,360	41	79	20	15,000–25,000
Galapagos Islands	7,845	136	25	38	0	

Source: Groombridge 1992; Wheatley 1994; Heywood 1995.

All five countries of Table 3.6 harbour tropical rainforest. For comparison Table 3.7 provides details of EBAs for Zimbabwe, Cameroon, Mauritius and UK.

Bird species diversity patterns can now be analysed using extensive DNA databases. Recent evidence suggests that many endemicity hot spots in South America and Africa may be a mix of ancient lineages that have survived and recent (Pliocene/ Pleistocene) radiations of other bird groups. The habitat heterogeneity and disturbance in the hot spots simultaneously allow relict species to cling on in some patches but also act as a spur to radiations in adjacent habitats. These species factories are often found in the zones between major biomes. (See Figure 3.10.)

Plant species diversity is unevenly divided between four main phyla: the Bryophytes (mosses and liverworts) 16,000 species; Pteridophytes (ferns and allies) 10,000–12,000 species; Coniferophytes 700 (Conifers, often called by slightly older name Gymnosperms); Anthophytes, 250,000, perhaps up to 750,000 (flowering plants, often called Angiosperms). Pteridophytes, Coniferophytes and Anthophytes are often separated out from Bryophytes as the vascular plants, defined by possession of effective water transporting tissues. Diversity within the three

Plate 12 *Mount Kupe, part of the Cameroon Mountain Endemic Bird Area*

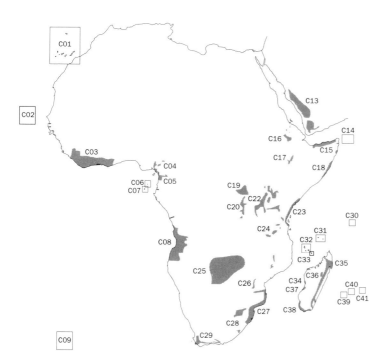

Figure 3.9 *African Endemic Bird Areas. The prefix 'C' refers to EBAs from Africa, Europe and the Middle East under the International Council for Bird Preservation EBA scheme. Mount Kupe (Chapter 4) is part of the Cameroon Mountain EBA, C04, Mauritius, once home to the Dodo, C40 and Zimbabwe harbours C26, the east Zimbabwean Mountain EBA*

Source: Redrawn from Bibby 1992.

vascular plant groups is biased towards some families. Anthophyte diversity is dominated by a few families, e.g. Orchidaceae (orchids) with between 25,000–35,000 species, Compositae (thistles, dandelions) 20,000 and legumes (peas) 14,200. The Coniferophytes contain 500 conifers, 100 cycads then some numerically tiny but remarkable species, notably the fossil relict Gingko tree (one species) and the slow-growing, two-leafed *Welwitschia* of southern African deserts. The Pteridophytes consist of 9,000 fern species and, again, small numbers of allied groups including horsetails and Lycopods, plants that were once the dominant vegetation of Palaeozoic forests growing as trees. Plants are unevenly distributed between the continents: Latin America 85,000 species; Africa 40,000–45,000; Asia 50,000: Australia 15,000; North America 17,000; Europe 12,500. Eighteen plant diversity hot spots support 50,000 species, 20 per cent of the world's vascular flora, but between them cover only 5 per cent of the land's area.

Species richness within the sometimes unnatural confines of political boundaries reflects size, topographic heterogeneity and complexity plus climate. Detailed studies of tree and vine inventories from South American tropical forests have compared species richness to climatic factors such as total rainfall and seasonality, plus soil nutrients. Diversity showed clear increases with total rainfall and length of rainy season. Although climate alone was a very effective predictor of species richness, soil factors could be important, especially increased species richness with increased nitrogen availability at higher altitudes. Such results can be useful to focus attention on known high total rainfall/ long rainy season sites as of particular importance for conservation, even if good species inventories are not yet available. The greatest species richness is in the tropical forests but many oceanic islands have

Table 3.6 *Five countries with largest numbers of Endemic Bird Areas*

Country	EBAs (inclusive of those shared with other countries)	Restricted range species confined to countries' EBAs	Total restricted range species, confined plus shared	Priority I EBAs	Priority II EBAs	Priority III EBAs
Indonesia	24	339	411	16	7	1
Brazil	19	122	201	8	8	3
Peru	18	106	216	4	7	7
Colombia	14	61	189	4	9	1
Papua New Guinea	12	82	172	3	63	3

Source: Bibby 1992.

very high endemicity. Tiny St Helena in the Atlantic, Napoleon's final exile, has 74 endemics out of 89 species. Table 3.8 gives known vascular plant numbers and endemicity for the top five species rich countries and other countries for comparison. (See Figure 3.11.)

Revelations of plant species richness continue at a smaller scale. A study of a one hectare plot in the Ecuador *terre firme* tropical forest in the late 1980s found 473 species (with stem diameter at standard breast height of greater than 5 cm). Given the estimated 3,000 tree species in Ecuador this represented 16 per cent of the country's tree flora but 49 per cent of species were only represented by one individual. Similar recent inventory work from Colombian forest, working in even smaller plots, 0.1 hectare area but including trees and herbs, found up to 313 species per plot.

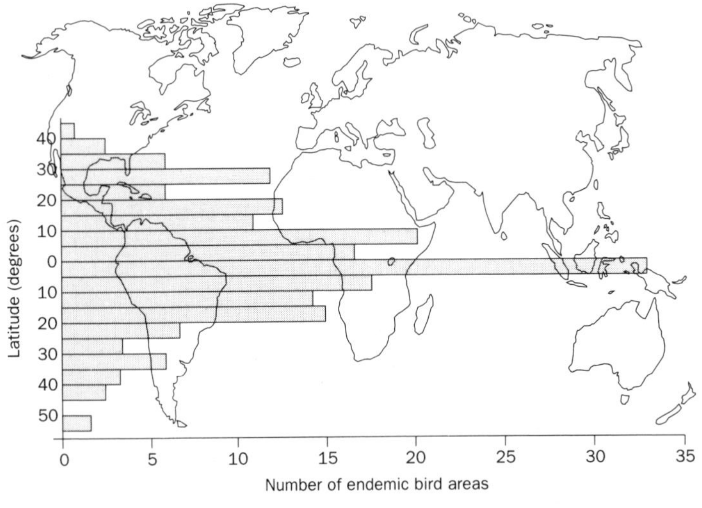

Plant species diversity has also been the target of the IUCN Plant Conservation Programme to identify Centres of Plant Diversity. The rationale for CPDs is that their identification and subsequent protection will conserve the majority of wild plants; 241 CPDs have been identified so far. The main criteria are species richness and numbers of endemics. The Programme also takes into account value of the gene pool to humans, diversity of habitat types,

Figure 3.10 *Latitudinal gradient of numbers of Endemic Bird Areas*
Source: Redrawn from Bibby 1992.

Table 3.7 Endemic Bird Areas of Zimbabwe, Cameroon, Mauritius and UK

Country and name of EBA	Area of EBA, km²	Restricted range species confined to EBA (+ shared with other EBA)	% of EBA area protected	Priority category
Zimbabwe East Zimbabwe Mountains	4,900	2 (+ 2)	5–10	III
Cameroon Cameroon Mountains	7,300	26 (+ 2)	5–10	I
Cameroon Cameroon/ Gabon Lowlands	40,000	5 (+ 1)	20–30	III
Mauritius Whole island system	1,900	6 (+ 4)	0–5	I
UK No EBAs	—	1	—	—

Source: Bibby 1992.

Table 3.8 Flowering plant diversity, selected countries

Country	Flowering plant species	Conifers and allies	Ferns and allies	Number of endemics (% of species)
Brazil	55,000	—	—	—
Columbia	35,000	—	—	1,500 (4.3)
China	30,000	200	2,000	18,000 (56)
Mexico	20,000–30,000	71	1,000	3,624 (13.9)
USSR (as was)	22,000	74	207	—
Madagascar	8,000–10,000	5	500	Up to 8,500 (68)
UK	1,550	3	70	16 (1)
Zimbabwe	4,200	6	234	95 (2.1)

Source: Groombridge 1992; Heywood 1995.

sites containing significant proportions of species adapted to specific conditions found therein and imminent large-scale threat.

Ecological biodiversity

Describing the extent of global ecosystems

Ecological biodiversity shows global patterns akin to taxonomic trends. Global classifications depict polar to equatorial biomes, largely created and defined by geography and climate. Some habitats are restricted by latitude, e.g. tropical rainforest, coral reefs, but others occur across the planet, though their precise regional form and complement of species varies.

Global biogeography has long recognised differences between the major continents and their allied peripheries. The six main divisions are Nearctic (North America) Neotropical (Central and South America), Palaearctic (Europe, Russia, central and eastern Asia), African (Africa including Madagascar), Oriental (India, Indo-China and South East Asian islands down to Borneo and Australasia (Australia, New Zealand, New Guinea). (See Figure 3.12.)

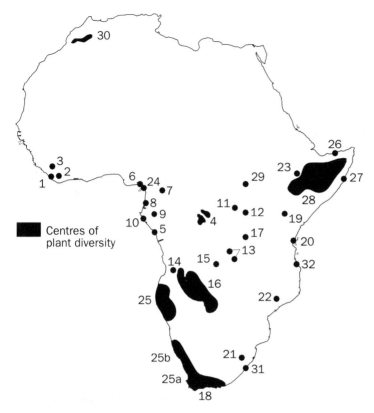

Figure 3.11 *African Centres of Plant Diversity identified by IUCN.*
Sites are selected due to particular richness of flora which, if
protected, would safeguard the majority of the earth's wild plants

Source: Redrawn from Groombridge 1992.

Increasingly detailed global classifications of biomes have been devised using remote sensing from satellites. These attempt to classify the main terrestrial ecosystem types producing schemes that broadly reflect global climate systems and sometimes other natural or human factors e.g. soils or landuse. Udvardy's 1975 combination of the use of dominant vegetation type and biogeographic is still widely employed and has been used as a base against which to measure human impacts on biomes at a global scale. (See Figure 3.13.) Udvardy based his system on eight major biogeographic realms: Afrotropical, Antarctic, Australian, Indo-Malayan, Nearctic, Neotropical, Oceanian and Palaearctic. Major vegetation types, biomes, were then assessed within each. The biomes were identified as provinces reflecting regional variations in form. Table 3.9 lists Afrotropical and Neotropical examples.

The distribution of wetlands varies globally, regionally and nationally. The extent of global wetlands has been estimated from compilations of maps and remotely sensed images from satellites. Estimates of the global area of wetlands vary with definitions and inclusion of coastal wetlands but range between 5.3 and 8.6 million sq km, compared to estimates for extant tropical rainforest of 9.4 to 12 million sq. km and for grasslands of 24 to 35 million sq. km. The total extent of wetlands varies with latitude and the different types of wetland are unevenly distributed with vast tracts of bogs dominant across northern temperate and subarctic continents versus the swamps and flood plains of the tropics.

The distribution varies with scale. Table 3.10 compares wetland distributions globally and in Europe. Differences would affect assessment of conservation priorities. (See Figure 3.14.)

The importance of global ecosystems: functional biodiversity

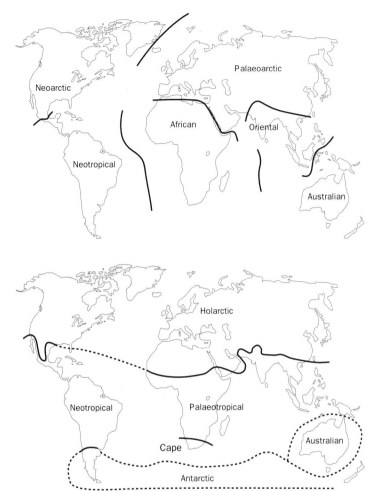

Figure 3.12 *The classical animal (above) and plant (below) global biogeographical realms*

Source: Redrawn from Heywood 1995.

Central to the concept of biodiversity is the sense that ecosystems are important for what they do. Ecosystems carry out functions, thereby providing services. Ecosystems are responsible for fluxes of energy and materials. The importance of ecosystems as providers of services adds to our awareness of the dangers from degradation and loss and of the value of biodiversity. Functional biodiversity meshes the concept of the ecosystem, dominated by ideas of flow and flux of resources with the community, the numbers of species. Ecosystems are a synergy of genetic, population, community, ecosystems and landscape biodiversity. Ecosystem function is the sum total of their activity, apparent even when the precise importance of individual species as keystones driving the processes remain unclear. The array of services includes vital global life support through atmospheric quality, climatic control, protection of coasts and nutrient cycling. The importance of functional biodiversity is best revealed by an example of just one ecosystem.

The importance of wetlands

Wetlands throughout the world support their own unique wildlife. By its very existence this biodiversity has an intrinsic value. Tangible, highly valuable benefits also derive from wetland ecosystems. These can be broadly divided into **production** (in ecological terms the living components, in economic, the stock), the **ecosystem functions** (in economics, the services) and the **intrinsic cultural and biological diversity** in its own right (attributes to economists). The astonishing variety of benefits, local to global, provided by wetlands is outlined next.

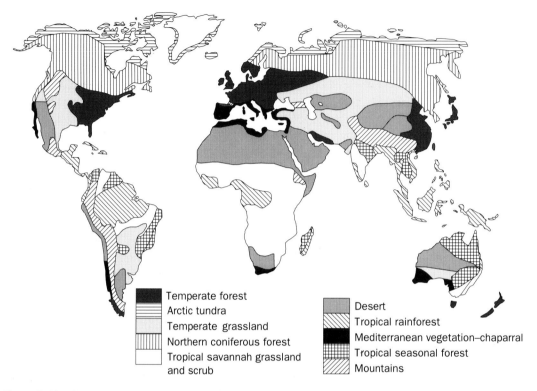

■ Temperate forest	▨ Desert
▤ Arctic tundra	▨ Tropical rainforest
▦ Temperate grassland	■ Mediterranean vegetation–chaparral
▥ Northern coniferous forest	▦ Tropical seasonal forest
☐ Tropical savannah grassland and scrub	▨ Mountains

Figure 3.13 *Gross global vegetation classification*

Source: Redrawn from Heywood 1995.

Production

Wetlands are the most productive natural ecosystems on Earth. The following examples are all sustainable uses of wetlands, though in some cases more intensive versions of the same exploitation also exist, destroying the original habitat. (See Plate 13.)

- Consumptive. Wetlands yield fish, shellfish and prawns. Plant products include wild fruit and vegetables, gardened crops and small-scale arable. In addition plants provide grazing plus harvested fodder, building poles, thatch, craft materials, latex, resin, tanning products, beer and medicines. Note that these products are not the preserve of subsistence economies. High vale crops, e.g. thatch, rushes, sedges and wild cranberries, are harvested in developed countries. Animals are hunted for fur and skins, as well as sport. Energy is provided by fuelwood and peat.
- Non-consumptive. Tourism and recreation rely on the wetland landscape and wildlife without being direct consumers. (See Plate 14.)

Services

- Storage. Wetlands act as major water stores, allowing ground water recharge as water slowly seeps into deep aquifers and also discharge, as ground water tops up surface aquatic habitats such as rivers and lakes. Note that wetlands can act as both recharge and discharge sources, perhaps switching roles with the seasons. Even small wetlands can act as insurance supplies during drought. Trapped sediment can build up land. Peat bogs are major sinks for carbon dioxide because peat is organic (primarily plant) material that cannot completely decay so that the carbon is not recycled.

- Buffering. Buffering is the ability to slow, compensate and ameliorate against change. Wetlands buffer many potentially destructive environmental processes, reducing both the size and rate of change. Coastal wetlands provide shore defences, dissipating wave and storm energy. Wetlands act as flood overspill areas holding water which may do damage elsewhere, slowing the rise of floodwaters and desynchronising floodwater peaks from different rivers which might otherwise combine. Climate is also buffered so that in hot weather wetlands act to cool the local climate and in the cold to warm up their environs. This benefit has been used in fruit farms in the USA to guard against sudden cold snaps.

- Cleaning. Wetlands are effective filters, particularly between rivers and their surrounding catchments. Materials are trapped and held, the wetlands acting as sinks, and rates of flow are slowed, allowing extra time for natural processes to neutralise potentially harmful inputs. Wetlands are known to trap and clean sediment run-off, organic detritus, sewage, excess nutrients (especially phosphorous and nitrogen compounds), acidic inputs, metals, pesticides and pathogens. These effluents can be from a readily identifiable point source such as a mine working or sewage treatment plant. In addition wetlands are particularly useful to control diffuse, non-point source inputs that plague so many catchments, e.g. nutrient run-off from farmland, which cannot easily be collected and cleaned by technical intervention.

- Pathways. The waterways ramifying throughout many wetlands provide pathways for natural fluxes such as nutrient recycling and alluvial deposition, increasing fertility. Open water also provides literal routes for movement by animals and humans.

Attributes

Cultures utilising wetlands have developed customs, architecture and landscapes as a result of this interdependence. In some cases the entire society is intimately tied to the wetland system, creating unique cultural systems, e.g. the Marsh Arabs.

So often dismissed as miasmal wastes in urgent need of draining, wetlands are actually a vital asset (see Box 12).

Islands and hot spots: the Earth's extra special places

Biodiversity is not evenly distributed. Within the general patterns of taxonomic and ecosystem distribution there are striking concentrations of species, associated with certain ecosystems. Identification of these special places is important for practical conservation and to understand the underlying causes that drive diversification. These foci can be special by virtue of several sometimes contrasting, factors:

Table 3.9 *Example vegetation types and extent from Udvardy's system*

Realm	Biome and province	Area, km²
Afrotropical	**Tropical Humid Forest**	
	Congo rainforest	2,195,019
	Guinean rainforest	709,112
	Malagasy rainforest	147,862
	Evergreen Sclerophyllous vegetation	
	Cape Sclerophyll	99,663
Neotropical	**Tropical Humid Forest**	
	Amazonian	2,864,623
	Campechean	279,695
	Colombian coastal	273,266
	Guyanan	1,090,396
	Mudieran	1,988,840
	Panamanian	128,872
	Atlantic forest	223,944
	Evergreen Sclerophyllous vegetation	
	Chilean sclerophyll	47,988

Source: Hannah *et al.* 1995.

- very high total species numbers;
- endemicity, whether of common or unusual lineages;
- unusual combinations, characteristics of communities;
- super speciose taxa.

These jewels in biodiversity's crown fall into three contrasting types. First, **continental hot spots** which are sites of very high diversity, often also with unusual endemics, sometimes called mega, hyper and super diversity centres. Second, there are **large islands** (sometimes called **continental islands**) harbouring diverse distinctive faunas which include relict fauna long extinct on the main continents. Finally, there are small **oceanic islands**, often low in total species numbers (though a few individual taxa can be unusually speciose) but with high proportions of endemics, unusual combinations of species, and peculiar evolutionary lineages. Merely listing high diversity centres by country is possible but not very informative since many, especially the continental centres, do not occupy the whole of a country and some straddle borders.

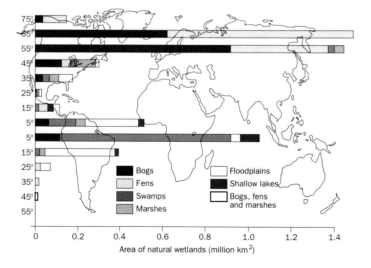

Figure 3.14 *Latitudinal extent of different wetland types*

Source: Redrawn from Aselmann and Crutzen 1989.

Table 3.10 *Extent of freshwater wetlands, globally and in Europe versus Africa (1,000 km)*

	Bogs	Fens	Swamps	Marsh	Floodplain	Lakes	Mangrove	Anthropogenic
Global	1,867	1,483	1,130	274	823	114	27	1,300 (rice paddies)
Europe	54	93	1	4	1	1	0	Minor
Africa	0–38		85	57	174	39	6	46

Source: Groombridge 1992.

Plate 13 *The value of wetlands: direct use production. Saw sedge (Cladium mariscus) harvested at Whicken Fen, Cambridgeshire. The sedge is used to protect peaks and angles of thatched roofs and is a high value crop*

Continental hot spots

Biodiversity hot spots defined by endemism and the uneven distribution of species have been cited for birds and plants in earlier sections. A few countries, primarily tropical, have been described as Megadiversity Countries, unusually rich in all forms of biodiversity, although data for such categorisation relies on higher vertebrates, plants and a few insect groups. The Megadiversity Countries are Mexico, Columbia, Ecuador, Peru, Brazil, Zaire, Madagascar, China, India, Malaysia, Indonesia and Australia. However, political boundaries are an inappropriate guide and generalisations do not take into account the uniqueness and biodisparity found in other countries. Eighteen global hot spots identified by endemicity of many taxa have been identified. Table 3.11 summarises data for four.

Globally significant continental hot spots are often regions of habitat heterogeneity caused by habitat change. The resultant patchwork allows older taxa to survive and radiation of new species. (See Figures 3.16 and 3.17.)

Islands

Island diversity centres are very different to the continental hot spots, often poor in species but sheltering unusual relics, strange combinations and extraordinary

Box 12

The role of wetlands. Global and national ecosystem services

Global: methane production

Methane, CH_4, is a greenhouse gas, also important in the formation of ozone. Concern at possible global warming has prompted detailed analysis of global biogeochemical cycles, including methane production which arises from natural sources such as digestive processes of ruminant animals such as cows and decay of organic matter, plus human sources such as use of fossil fuels and burning. Atmospheric methane is increasing by about 1 per cent per year, i.e. 40–50 Teragrams (1 Teragram = 10^9 kg) which, allowing for the balance between production, recycling and use, requires about 400–600 Teragrams to be pumped into the atmosphere annually. Estimates of methane production from wetlands are 40–160 Teragrams per year from natural wetlands, plus 60–140 Teragrams from rice paddies. The most important regions are northern latitude fens between 50-70°N, subtropical paddies between 0–20°N and southern hemisphere tropical swamps between 0–10°S. Although methane emissions from wetlands are highly variable depending on seasonal temperature and waterlogging and estimates of global methane fluxes are still tentative, the suggested mean methane emissions from wetlands may represent up to half the global annual production of methane and are therefore very important for planetary health.

National: dambo horticulture in Zimbabwe

Small seasonally inundated valley wetlands, sometimes herbaceous, sometimes wooded, are found in the headwaters of drainage basins throughout southern Africa. Zimbabwe, although primarily an arid country, is particularly rich in these ecosystems, known by the local name of dambos. Zimbabwean dambos have been used since at least the Iron Age as part of a shifting system of cultivation and grazing. Dambos are fertile and remain moist even in years with poor rainfall so they are an especially valuable safety net in an uncertain world. Dambo wetlands are primarily used as small (0.1–2 hectare) market gardens. The fertility and water supply permit diverse crops from staples such as maize and rice to water greedy pumpkins and fruit, which can be cropped all year. Limited grazing and beekeeping are possible. Dambos also act as water sources for drinking, rivers, irrigation and livestock during drought and as water stores, mopping up excess during rare floods. Dambos are a special asset to the rural poor as the horticultural garden produce is a valuable source of income. The nature of this local, small scale horticulture is a valuable economic opportunity for rural women. During recent drought years 80 per cent of households with access to market gardens on dambo wetlands remained self-sufficient in food. (See Figure 3.15.)

evolutionary radiations. Islands also have a historic role in our awareness and understanding of biodiversity, whether the evolutionary inspiration of the Galapagos Islands, or that icon of extinction, the Dodo, from Mauritius.

Islands have distinctive ecological communities. Historical contingency (what was living there before islands split from mainlands, when the separation occurred) and the vagaries of colonisation create unusual combinations of wildlife. Relict faunas long extinct on main continents survive and in some cases flourish. In situ speciation creates spectacular radiations, such as the very speciose fruit flies of Hawaii. Animals take up unexpected roles; on Hawaii one moth has a caterpillar that does not chew leaves but is a deadly ambush predator, snatching other insects. A few taxa

Table 3.11 *Example of continental biodiversity hot spots*

	Higher plant species	Endemic mammal species	Endemic reptile species	Endemic amphibian species	Endemic swallowtail butterfly species
Cape region, South Africa	6,000	15	43	23	0
Colombian Choco	2,500	8	137	111	0
Western Ghats, India	1,600	7	91	84	5
South West Australia	2,830	10	25	22	0

Source: Bibby 1992.

have become particularly characteristic of islands, e.g. pigeons, rails and tortoises, all capable of erratic but long-distance dispersal, yet not so mobile that once established new arrivals dilute the evolutionary aftermath. The same traits develop among these taxa: gigantism in birds and reptiles; dwarfism in mammals; flightlessness in birds. The wildlife of islands is also distinguished by being extinction prone.

There are two main types of high diversity islands, continental islands such as New Zealand, Australia and Madagascar and tropical oceanic islands, e.g. Hawaii, Galapagos and Mauritius.

Plate 14 *The value of wetlands: ecosystem services. The Insh marshes in Scotland, a Royal Society for the Protection of Birds reserve, act as a buffer, holding flood waters that spill out over the marshes from the River Tay, helping to protect land downstream from flooding*

Continental islands

Table 3.12 gives comparative biodiversity data for Australia, New Zealand, Madagascar and Kenya as a contrasting continental country.

Oceanic islands

Oceanic islands supporting special diversity are typically very isolated, volcanic (the altitude creating diverse habitats) and tropical. Low-lying tropical atolls lack habitat diversity while temperate islands are often too exposed to extremes of weather. (See Table 3.13).

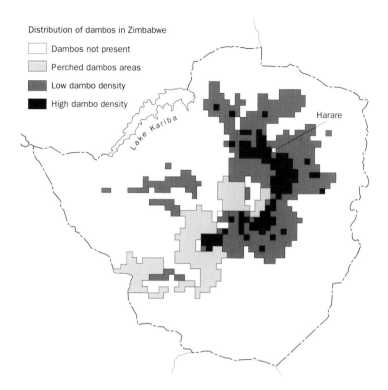

Distribution of dambos in Zimbabwe

☐ Dambos not present
▨ Perched dambos areas
▨ Low dambo density
■ High dambo density

Lake Kariba

Harare

Figure 3.15 *Extent of dambos in Zimbabwe*

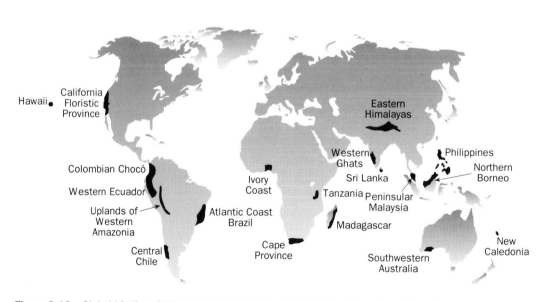

Hawaii•

California Floristic Province

Eastern Himalayas

Colombian Chocó

Western Ecuador

Uplands of Western Amazonia

Central Chile

Ivory Coast

Atlantic Coast Brazil

Cape Province

Western Ghats

Sri Lanka

Tanzania

Peninsular Malaysia

Madagascar

Southwestern Australia

Philippines

Northern Borneo

New Caledonia

Figure 3.16 *Global biodiversity hotspots, defined by endemicity and also threat from human pressures, identified by Wilson 1992*

Source: Redrawn from Wilson (1992).

Table 3.12 *Diversity, threat and uniqueness of three continental islands, with Kenya as a comparison. Note that counts of extinct species exclude those lost from very small islands around the mainlands*

	Australia		Madagascar		New Zealand		Kenya	
	Total	*Endemic*	*Total*	*Endemic*	*Total*	*Endemic*	*Total*	*Endemic*
Flowers			8,000–10,000		2,160		6,000	
Gymnosperms	15,000	80%	5	68%	22	82%	6	4%
Ferns			500		189		600	
Mammals	282	210	105	67	—	3	309	10
Birds	571	351	250	97	285	74	1,067	7
Reptiles	700	616	252	231	40	40	187	15
Amphibians	180	169	144	142	3	3	88	10
Fish	113	110	40	38	30	27		
IUCN Plant Centres of Diversity	8		1		3		1	
Endemic Bird Centres	7		5		2		2	
Threatened species								
Plants	2,024		194		232		144	
Mammals	38		50		1		17	
Birds	39		28		26		18	
Reptiles	9		10		1		2	
Amphibians	3		0		3		0	
Fish	16		0		2		0	
Animals known to have gone extinct	18		3		14		0	
Extinct endemic plants	173		0		5		0	
Unusual wildlife	Marsupial and reptile diversity		Lemurs. Extinct flightless Elephant Bird		Flightless birds (Kiwi, extinct Moa)			

Source: Groombridge 1992.

Island wildlife has proven particularly susceptible to extinction, due to several factors.

- Evolutionary innocence. Island species have proven very vulnerable to introduced competitors, predators, parasites and diseases. Many islands lack taxa such as large mammalian predators and ants, the very groups that are dominant ecological keystone taxa on mainlands. Relict fauna and particular traits such as flightlessness further increase the threat posed by alien arrivals.

Table 3.13 *Diversity, threat and uniqueness of three oceanic island systems. Note that the data exclude introduced species*

	Hawaiian Islands (Pacific)	Galapagos Islands (Pacific)	Mauritius (Indian Ocean)
Vascular plants	900 of which 850 are endemic	540 of which 170 are endemic	878 of which 329 are endemic
Threatened native plants	343	82	269
Introduced plant species	4,000	195	—
IUCN Plant Centres of Diversity	1, whole archipelago	1, whole archipelago	1, whole archipelago
Endemic Bird Centres	1, whole archipelago	1, whole archipelago	1, whole archipelago
Animals known to have gone extinct	86	5	41
Extinct endemic plants	108	2	24
Species of land snails	c 1,000, all endemic	90 of which 66 are endemic	109 of which 77 are endemic
Extinct snail species	29	1	25
Unusual wildlife	High endemism, super-species rich taxa, e.g. fruit flies	High endemism, unusual species, e.g. marine iguana	High endemism, once home to the Dodo

Source: Stone and Stone 1989; Groombridge 1992.

- Small populations. Many island species occur as small, isolated populations specialised to live in a narrow habitat range and with no pool from which recolonisation can boost numbers. Genetic diversity may be enfeebled by inbreeding and bottlenecks due to population size. In addition coevolution of island taxa one with another may result in extinctions cascading through the island once one or two species are lost and ecological links are broken.
- Lack of disturbance. Physical disturbance to island habitats can be increased once humans arrive, e.g. direct action such as fire to clear land or indirectly through impacts of introduced animals such as pigs grubbing up land.
- Human exploitation. Direct human harvesting has wiped out species such as the Dodo and Moas and endangered others, e.g. Giant Tortoises. Habitat loss as humans clear land is just as great a threat, especially given the small habitat areas and species' ranges of many island taxa.

Islands remain under threat, even the most famous. The Galapogos have become a recent focus of concern, in part due to increased tourism but also due to protests from local people who are angry at restrictions due to conservation laws. Protests have culminated in intentional habitat destruction and threats to conservationists.

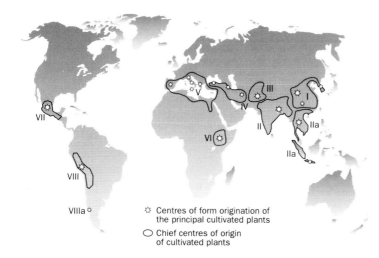

Figure 3.17 *Centres of origin of cultivated plants, based on the work of Vavilov. Stars indicate the centres of origination for the form of crop plant used, the outlined surrounding areas the broad region from within which the plant was domesticated*

Source: Redrawn from Heywood 1995.

More things in Heaven and Earth . . .

Fragments of ancient continents, far flung islands and impenetrable jungle – our fascination with biodiversity is in part the mystery and unknown. Sad to say, despite recent expeditions there is unlikely to be a population of Brontosaurs lurking up the River Congo but there are other surprises. Mauritius, which is now recognised as an oceanic hot spot, was the home of the Dodo, *Raphus cucullatus*, discovered in 1598 and extinct by 1670. The Dodo was the epitome of the absurd, a fat waddling flightless pigeon permanently clad in juvenile plumage, too stupid to avoid being eaten into extinction. Recent analysis of fossil bone structures reveal an athletic leggy creature confirmed by the earliest illustrations. The Dodo was an innovative design but fatally maladapted to human interference.

Continental hot spots also hold surprises. Several virulent diseases, Lassa fever and Ebola virus, seem to flare from African rainforest hot spots. Ebola outbreaks not only kill humans. In 1994 a 40-strong Chimpanzee clan (*Pan troglodytes*) in the Tai Forest of the Ivory Coast lost 12 to Ebola. Ebola virus replication is error prone; the frequent mutations that result permit infection of a variety of species. The Tai chimps were proficient hunters of small mammals and perhaps picked up the disease from the rodents that have boomed since 1990 when Liberian refugees from civil war crossed into the area. The Dodo and the Tai chimps may seem idiosyncratic but they prompt important questions. How do human pressures, whether hungry sailors or refugees, impact natural systems? Chapter 4 describes species extinctions, ecosystem loss and the underlying causes for the degradation of biodiversity.

Summary

- Biodiversity is categorised as genetic, organismal and ecological. Domestic and cultural categories can be added.

- Measures of the richess of extent of any category remain fraught. Estimates of total species alive today range between 10 and 30 million.

● Functional biodiversity is very important to planetary health. Ecosystems function, as a result of species activities, providing services to the wider environment.

Discussion points

1 Is the Giant Panda more important than the Polio virus?

2 If all the wetlands in your country were gone what would have been lost?

3 The biodiversity of islands can be strange, special and frightening. Why?

See also

Role and value of biodiversity, Chapters 1 and 4.
Islands and extinction, Chapter 4.
Measuring biodiversity loss rates, Chapter 4.

General further reading

'Magnitude and distribution of biodiversity'. D. L. Hawksworth and M. T. Kalin-Arroyo (eds). 1995. In V. H. Heywood (ed.). *Global Biodiversity Assessment*. Section 3. CUP for UNEP, Cambridge.

Global Biodiversity. Status of the Earth's Living Resources. B. Groombridge (ed.). (1992). Chapman and Hall, London.

Both books are essential and contain detailed inventories of biodiversity patterns.

The Diversity of Life. Edward O. Wilson. 1992. Harvard University Press.
Reviews extent of biodiversity, especially importance of hotspots.

Putting Biodiversity on the Map. C. J. Bibby (ed.). 1992.
International Council for Bird Preservation, Cambridge.

'How many species are there on earth?' R. M. May. 1988. *Science*, vol. 241, 1441–1449.
Robert May reviews revived interest in this question prompted by rise of biodiversity research in 1980s.

④ Extinction

Extinction and habitat loss epitomise our sense of biodiversity crises. This chapter covers:

- **Extinction rates and ecosystem loss**
- **Causes of extinction**
- **Human pressures on biodiversity**
- **Valuing biodiversity.**

Extinction can bring with it all the familiarity and fame akin to that of a dead pop star. The Dodo and dinosaurs are household names, evocative of failure. Neither deserve this epitaph and the confusion between types of extinction, natural versus anthropogenic, hinders our understanding of the rates and causes of loss. This chapter describes losses to biodiversity during the current crisis, driven by human activities, the causes, both ecological and economic and the consequences.

Current losses of biodiversity

Prophets and loss: estimating current extinction rates

The fate of all species is extinction, perhaps leaving evolutionary descendants as a lingering echo. Extinction, whether background or mass, is a natural process unleashing evolutionary creativity. Quantitative and qualitative changes to extinction rates define the current biodiversity crisis. However, the growing awareness of extinction, fuelled by slogans such as 'extinction is forever', has proved difficult to quantify. Several approaches have been used to measure recent extinction rates and to project future trends.

Confirmed extinctions

The precise start of the current crisis, defined by the impact of the human, is hard to delimit. Unusual bursts of extinction coincided with human arrival in Australia (30,000–50,000 years ago), North and South America (11,000–12,000 ya), Madagascar (1,400 ya) and New Zealand (1,000 ya). In every case the same types of animals were disproportionately affected, with severe losses of large mammals

Box 13

The Clovis overkill extinctions

A wave of extinctions hit North American mammals between 11,000–12,000 years ago. Over the comparatively short span of 1,000 to 2,000 years at least 335 species were lost from small insectivores to largest elephants. These losses continued down into South America. Charles Darwin witnessed the destruction among fossils, lamenting 'now we find mere pygmies compared to the antecedent allied races'. Losses of large mammals, the megafauna, are especially dramatic, particularly of endemic American taxa (90 per cent of herbivore megafauna) while invading species fared much better. The losses are coincident with climatic changes but peculiar in their selectivity of large native species and progress across the continent. These extinctions have been linked to the spread of humans, invading via a land bridge from Asia and reaching what is now the southern USA by 11,000 years ago. These people are often dubbed the Clovis culture. The Clovis invasion may have brought in new hunting techniques, with the hunters butchering their way south. The naive native megafauna were hopelessly vulnerable and extermination of these prey species lead to losses among native carnivores. Mastodons, mammoths and giant ground sloths were wiped out while invading mammals which spread in from Asia alongside the hunters and were used to human pressure survived. The Clovis overkill remains contentious. Perhaps a combination of climatic first strike and human hunter follow-up combined to destroy the giant mammals. (See Figure 4.1.)

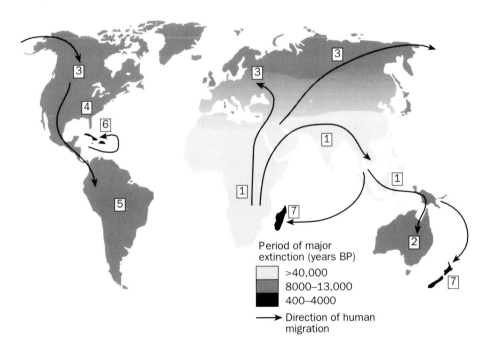

Period of major
extinction (years BP)

	>40,000
	8000–13,000
	400–4000

⟶ Direction of human
migration

Figure 4.1 *Waves of major extinctions in the wake of humanity's advance across the planet*

Source: Redrawn from Martin P.S. Prehistoric overkill. In Martin P.S. and Wright H.E. jr. (eds) (1967) *Pleistocene Extinctions*. Yale University Press, New York.

(megafauna) and flightless birds. This characteristic pattern suggests a similar cause active on different continents at different times. In North America the losses have been tracked as an advancing wave coincident with the spread of early indigenous people down the continent and dubbed the Overkill Theory (Box 13). No such losses are evident in Africa, which may refute human impacts as a cause or may reflect the much longer coevolution of humans and African wildlife.

The evidence of megafauna overkill by people is circumstantial but throughout the last 40,000 years waves of extinction have hit continents and islands coincident with human arrival. Coincident climatic change may explain the American and Australian losses with humans as the final blow to populations already stressed and fragmented by habitat change. This pattern echoes Raup's idea of mass extinctions due to a first strike, weakening natural systems, with follow-up stresses mopping up vulnerable taxa en masse. The oceanic island losses are more convincingly due to humans. Losses occurred before European colonisation, the Moa and twenty other land birds lost to the Maoris in New Zealand, while in Hawaii analysis of bone deposits has identified 35 to 55 birds extinct during the Polynesian colonisation. Not only the birds lost out. Evidence from the Pitcairn Islands suggests that humans abandoned the islands once the fauna had been depleted. Some Hawaiian islands even contain abandoned prehistoric settlements with evidence of bird cults, culture that collapsed with the faunal extinctions. Recent estimates of total bird species losses from Pacific islands are at least 2,000 species, 20 per cent of all known bird species. (See Figure 4.2.)

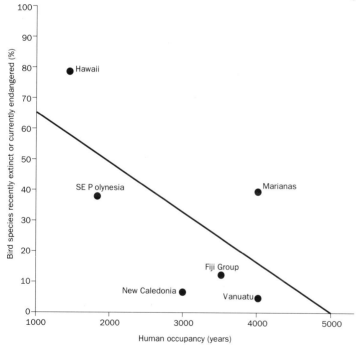

Figure 4.2 **Birds recently extinct and endangered on Pacific islands relative to length of human occupation. The numbers of species in immediate danger are highest on islands most recently occupied. Species vulnerable to human pressures ('wimp species') have long been lost from islands occupied earlier. The increased threat to the Marianas bird fauna is due to recent colonisation of the islands by the brown tree snake**

Source: Redrawn from Lawton and May 1995.

Documented extinctions have typically been recorded from 1600, following in the wake of European global exploration and settlement. Recorded extinctions are taxonomically biased to mammals and birds, with molluscs an unexpected third category due to intense work on some island species. Box 14 is a requiem for a very recent and precisely recorded

Box 14

A documented extinction, Partula turgida, *a snail*

At 17.30 hours on the 1 January 1996 the last individual of the snail *Partula turgida* from French Polynesia died in London Zoo. To time an extinction so precisely is very unusual. The snail was rescued from the wild in 1991 as part of a conservation programme for the endemic snails of Pacific islands set up in 1987, now nurturing 33 taxa dispersed between 18 captive stock refuges. Many Pacific island endemic snails have been driven to actual or imminent extinction by release of a predatory snail, *Euglandinia rosea*. This fast moving mollusc was introduced as a would-be biocontrol for giant African land snails, *Achatina fulica*. African snails were imported as food. They escaped and flourished, becoming a pest species attacking crops. *Partula* species have aroused attention not simply because of documented losses, detectable from remnant shells, but from their fame as examples of evolutionary diversification, with some species apparently limited to individual valleys on some islands. In 1994 three species of *Partula* were reintroduced into a small enclosure in Moorea from captive breeding programmes in the UK. The walled enclosure was further defended by electric fencing. Unfortunately monitoring in 1995 showed that *Euglandinia* had got in by using fallen vegetation to bridge the barriers and had killed all the snails. More encouragingly shells of young snails showed the *Partula* were capable of breeding following release. The work continues. (See Plate 15.)

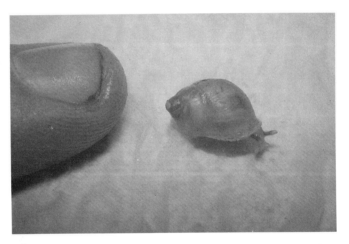

Plate 15 Partula turgida. *The last individual of this species before its extinction in 1996. London Zoo is home to captive populations of several* **Partula** *species, rescued from imminent extinction in the wild and now breeding in captivity. Their offspring are the foundation of reintroduction programmes*

Source: London Zoo. Photograph Dave Clarke.

species extinction. There is a geographical bias in extinctions towards Europe, North America and island systems, though the latter reflect a genuine vulnerability of island faunas (Table 4.1). Some supposedly extinct species cling on and have been rediscovered, so called Lazarus taxa, e.g. the Mount Kupe bush shrike featured later in this chapter. Some extant species number so few individuals that

they are doomed to extinction, dubbed living dead taxa, such as the poor *Partula turgida* of Box 14. Some wild populations are just as surely doomed. Their habitats are undermined, but they are clinging on simply because it takes time for ecological collapse and death of individuals to take place. (See Figure 4.3.)

While losses of mammals, birds and molluscs have been fairly well documented, scarcely 60 insect species are known to have become extinct and of these half are Hawaiian examples. (See Figure 4.4.)

Predicting extinction rates

Estimates of extinction rates rely on extrapolating data for known patterns of loss, much as estimates of total diversity rely on expansions of models. Five approaches have been used.

- estimates from recent past extinction rates;
- predictions from habitat loss;
- predictions fron changing status of threatened taxa;
- molecular;
- energy use.

Estimates from recent past extinction rates

Estimates depend on the quality of data. Extinction rates from fossil records are best documented for marine molluscs, not widely studied today, at least in the context of extinctions and conservation. The taxa arousing greatest concern recently, birds and mammals, have a very patchy fossil record. Fossil data suggest mammal extinctions at a rate of one species every two hundred years, compared to known losses of at least

Figure 4.3 *The dodo. Apocryphal symbol of extinction as doomed, dumpy convenience food but recently revealed as a lithe ground dweller, probably well suited to its island habitat but not to the arrival of humans*

Source: Redrawn from *New Scientist*, 28 August 1993.

Table 4.1 *Known extinctions up to 1995*

	Total	On islands
Mammals	58	34
Birds	115	104
Molluscs	191	151
Other animals	120	74
Higher plants	517	196

Source: Groombridge 1992.

twenty species this century. For birds extinction rates in the last 2,000 years are a thousandfold increase on loss rates prior to major human impacts. Estimates are also biased by the number of recorded extinctions from oceanic islands in the Pacific. Extrapolation of data from Pacific islands to, for example, continental Africa is likely to be foolish. For birds and mammals combined, a general estimate of extinction rates is a 1 per cent loss this century.

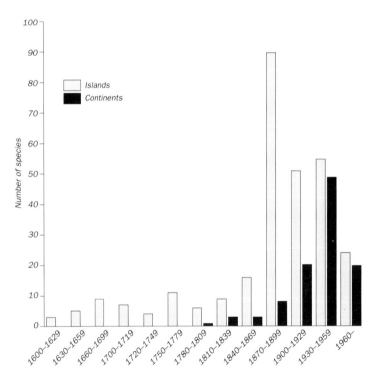

Figure 4.4 *Known animal extinctions from islands and continents since 1600*

Source: Redrawn from Heywood 1995.

Predictions from habitat loss

Ecological science provides a host of examples of species–area relationships (see Chapter 2). Essentially the larger the area of habitat, the more species it supports. Species–area studies derived primarily from work on oceanic islands but hold for any habitat that exists as isolated patches. As a gross rule of thumb, a 90 per cent decrease in area results in a 50 per cent loss of species.

One of the main causes of extinction is habitat loss. As the area of habitat declines species–area relationships predict that species numbers will decline. For habitats where loss rates have been measured and species–area patterns are known extinctions can be predicted. This technique has been used repeatedly to predict extinctions due to loss of tropical rainforests. Lovejoy's 1980 paper, one of the conceptual origins of biodiversity, uses this very approach. Estimates based on habitat loss for rainforest species vary with assumptions of forest loss rates, species richness in proportion to area, and variations between precise type of forest used for original data and species–area model used. Estimates of losses over the next twenty-five years vary between 2 per cent and 25 per cent, depending on taxa. Estimates of forest loss give a general prediction of extinction of between 1 and 10 per cent of all the world's species over the next twenty-five years, a 1,000–10,000-fold increase on background extinction rates. (See Figure 4.5.)

These projections have the advantage of including unknown undescribed fauna, albeit widely varied guesstimates. However, there are problems with this technique.

- Data and models. Species–area relationships have not worked so well for fragments of continental habitat as for real islands. Islands and continents may work to slightly different ecological rules. Known losses do not tally with model predictions. There is a disparity of estimated loss versus known species life spans. In part this is solved because some species though still alive are effectively doomed, they are living dead species.
- Habitat loss. Habitat loss is not evenly distributed so that predicting global losses from one habitat type cannot be accurate. Such predictions also ignore dangers to biodiversity hot spots and endemics. Estimates of habitat loss rates vary and can be poor. Such estimates assume loss of habitat is total. Habitat loss may not result in complete loss of species. Some species survive in replacement habitat.
- Scale. Species–area predictions use a simple correlation of species number to habitat area and assume loss rates will remain constant as habitat is lost. Extinction rates may accelerate as minimum habitat size thresholds are exceeded. Once fragments fall below a critical size they are effectively useless as habitat for some species, even though some remains. Habitat loss may also affect ecosystems function so that decreasing area causes cascading damage and disruption over and above linear predictions from the species–area relationship.

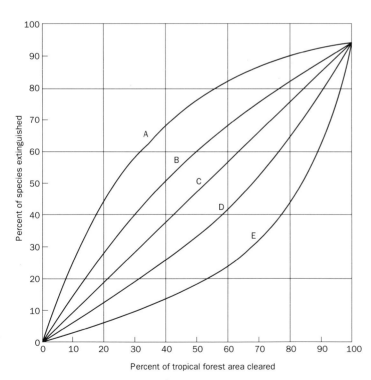

Figure 4.5 *Different models of species loss with forest clearance discussed by Lovejoy in his seminal 1980 paper*

Source: Redrawn from Lovejoy 1980.

Predictions from the changing status of threatened taxa

The International Union for the Conservation of Nature (IUCN) maintains internationally recognised lists of threatened taxa, the **Red Lists** and **Red Data Books**. These categorise taxa according to the intensity of the threat and include acknowledged extinct species. The rates at which species are added to these lists or move between categories can be used to estimate extinction rates.

Between 1986 and 1990 the animal species list has increased by more than 30 per cent but only 15 vertebrates and 33 animals in total have been added to the extinct category. Of plant species 163 have been added to the lists. For most taxa the lack of data for most species makes this approach poor, but for birds and mammals gives predictions of loss of half the extant species within 200 to 300 years, a time to extinction for half the extant vertebrate species of 7,000 years, and 3,000 years to wipe out half the known plant species.

Lawton and May (1995) note the coincidence of these techniques with those derived from species–area estimates for bird and mammal losses. There is general agreement. For birds, mammals and perhaps other terrestrial vertebrates current estimates of loss rates predict 50 per cent of species alive today will be lost within 200 to 500 years.

Molecular

Phylogenetic relationships can be explored at the molecular level, though the technology for detailed and extensive investigations has only recently been available. Molecular data from today's species give clues to the past, present and maybe future of a taxon. The story of a taxon's evolutionary history can be unravelled, including estimates of extinction rates that find an echo in the molecular baggage. Molecular phylogenies define individual lineages within larger taxa and the times of splits.

Analysis of phylogenetic trees defined by such precise molecular information has shown that extinction rates can be estimated for some taxa by comparing observed data to predictions of lineage history based on a model that supposes the probability of a new lineage starting or ending (extinction) is equal and constant throughout the taxon's history.

Energy use

Extinctions can be correlated with measures of human energy consumption, a surrogate measure for the pressure of human population growth and activity, fundamental factors driving species and habitat loss. Known species extinctions can be measured against known increases in energy consumption and future extinctions predicted by extrapolation against estimates of future energy consumption. One general estimate using this technique, based on energy increase predicted by the Brundtland Report, suggests that the time to extinction for half the remaining bird and mammal species is only 30 years. Comparison of known species extinctions of taxa, such as butterflies, from countries with reliable measures of energy use this century do not match the predicted loss rates form energy models. But energy-based models may correlate well with losses of scattered populations, reducing ranges and total populations.

Ecosystem loss rates

Loss of habitats has become a touchstone for conservationists. 'An area equivalent to the size of Wales' has become almost an SI unit for measurement of habitat destruction. Habitat loss deserves attention as both a measure of biodiversity loss and as a primary cause of species extinction. Monitoring habitat loss is surprisingly tricky. Precise definitions of and boundaries to ecosystems are hard to draw. Ecosystems are dynamic and their precise size, time scale of change and threshold beyond which they have been irredeemably wrecked are imprecise. Habitats can alter from one type to another; while a species is lost a habitat may become another habitat.

There are few reliable measures of global ecosystem loss rates other than for tropical rainforests. Loss of tropical forests has attracted special attention because they harbour the greatest species diversity. Estimates vary with precise definition of forest types (e.g. closed versus open forest) and as better coverage of more countries has been added to the database. Global percentage losses in recent years have been estimated as 1976–1980 0.6 per cent; 1981–85 0.62 per cent; 1981–90 0.9 per cent. Losses vary between countries. High losses, between 1981–8, include Costa Rica (4.0–7.6 per cent), Thailand (2.4–2.5 per cent) and Brazil (0.4–2.2 per cent). Losses are not solely a phenomenon of recent decades. Estimates of forest loss in Africa across recent centuries are 96,000 to 226,000 sq. km pre 1650; 470,000 ksq. m mid-nineteenth century to 1978, 51,000 sq. km 1981 to 1990.

Udvardy's global ecosystem classification (Chapter 3) has been used as a basis to assess loss and degradation of ecosystems at a global scale. Major vegetation types, biomes, were then assessed within each biogeographical realm. The biomes were identified as provinces reflecting regional variations in form. Table 4.2 lists Afrotropical and Neotropical examples, to compare with Table 3.9.

Table 4.2 Loss rates of example ecosystems, based on Udvardy's global classification. The percentage totally undisturbed is given, along with an index based on % partially disturbed or dominated by humans

Realm	Biome and province	Area, km^2	% totally undisturbed	Index of damage (0 = total loss, 100 = intact)
Afrotropical	**Tropical Humid Forest**			
	Congo rainforest	2,195,019	61.2	66.6
	Guinean rainforest	709,112	8.4	13.3
	Malagasy rainforest	147,862	39.4	39.4
	Evergreen Sclerophyllous vegetation			
	Cape Sclerophyll	99,663	13.5	18.5
Neotropical	**Tropical Humid Forest**			
	Amazonian	2,864,623	98	98
	Campechean	279,695	33.6	35.7
	Colombian coastal	273,266	45.9	52.3
	Guyanan	1,090,396	94.5	95.1
	Mudieran	1,988,840	81.5	82
	Panamanian	128,872	90	90
	Atlantic forest	223,944	6.5	12.6
	Evergreen Sclerophyllous vegetation			
	Chilean sclerophyll	47,988	36.1	45.8

Source: Hannah, Carr and Lankerani 1995.

Table 4.3 *Original and remaining forest habitat of different African countries, extent measured as km²*

	Approximate original extent	Estimated remnant	% remaining
Congo	342,000	213,000	62.4
Cameroon	376,900	155,000–179,000	41–48
Madagascar	275,000	41,700–103,000	15–37
Mauritius	1,850	30	1.6
Zimbabwe	7,700	80–200	1.0–26

Source: Groombridge 1992.

In the developed world we routinely and naively lump together disparate tropical countries on the assumption that they are all alike. Table 4.3 gives figures for forest losses in African countries cited as examples elsewhere in this book which show very different patterns of forest contraction.

Defining the endangered

The sheer variety of life on Earth inspires awe. Equally powerful is our concern for the endangered and rare. Animals sporting such labels are the familiars of umpteen television documentaries and public campaigns yet these terms can be bandied about with no precise definition. The California condor (*Gymnogyps californianus*) is rare, there are only a few individuals alive. Cuvier's beaked whale (*Ziphus cavirostris*) is rare, only a few have been seen off the Pacific and Atlantic coasts of Canada but there are more, perhaps many more, out in the wild. The scarce emerald damselfly (*Lestes dryas*) is rare in the UK, thought extinct in the 1970s and rediscovered at a few sites in the 1980s, but widespread across continental Europe and Asia. Precise definitions are not merely the realm of experts but are important for practical purposes, to raise awareness and to define the current threat to a species survival. Changes in status can be monitored and priorities for action decided.

Gaston (1994) reviewed the concept of rarity itself and showed that definitions varied with taxa, measures of abundance and extent of range. Population abundance or spatial range could be used quite independently. Precise thresholds (how small a population, how limited a range) go undefined. He concluded that many studies classify a certain number of species as rare because that is the number which feels right. Low abundance and/or small range define rarity, though the two often go together, and Gaston suggested a practical threshold for classification as rare as the 25 per cent of species in a taxon with the lowest abundance or range size. More abundant species can be described as common, those with a wider range as widespread. Even so, stating the scale at which judgement of rarity is made is an important part of any categorisation.

The term 'threatened' is widely recognised as an inclusive label for different degrees of danger to species' survival. In the 1960s the IUCN started to compile the Red Data Books, cataloguing known threatened species from around the world. The books have now spawned the Red Lists as the quantity of data grows, updated every two years by the IUCN and WCMC using advice from specialist groups. The Red Data Book approach has been widely adopted by individual countries (UK example) even regions

(UK county). Note that scale is important. Species may switch between categories from IUCN global lists to national and local Red Data Books. Box 15 outlines the IUCN categories.

Box 15

Red Data categories

In 1994 a revised ten category classification system replaced the previous six category system (extinct, endangered, vulnerable, rare, indeterminate and insufficiently known). The new categories are:

Extinct. No reasonable doubt that the last individual has died.

Extinct in wild. Only known to survive in captivity or naturalised well outside past range.

Critically endangered. Extremely high risk of extinction in the wild in immediate future.

Endangered. High risk of extinction in wild in near future.

Vulnerable. High risk of extinction in wild in medium-term future.

Conservation dependent. Taxon dependent on conservation programme which if stopped would place taxon into one of above categories within five years.

Near threatened. Taxon close to qualifying for one of above categories.

Least concern. Taxon does not qualify for any of above criteria.

Data deficient. Data insufficient to categorise taxon but listing highlights requirement for research, perhaps acknowledging suspicion that taxon warrants classification.

Not evaluated. Taxon not assessed.

Problems inevitably arise with such classifications.

- Bias. The categories only contain known species. The undiscovered plus species found but not yet described are omitted. The great majority of species is therefore excluded and the published lists could even divert attention from this unknown majority. Coverage for better known taxa varies: birds nearly 100 per cent; mammals 50 per cent; reptiles 20 per cent; amphibians 10 per cent; fish 5 per cent.
- Species based. The Red Data Books deal with individual species. There is no attempt to assess the threat to higher taxonomic levels which may represent more fundamental differences and variety or specialness.
- Lack of objectivity. The superficially neat, objective classification is based on expert opinion and therefore subjective. Risk is poorly quantified, e.g. percentage risk of extinction over a defined time period.
- Some criteria are not linked to threat. Red Data classification can be based on protection accorded to a species or unhelpful criteria, e.g. 'insufficiently known' (which applies to most life on the planet). Other potentially useful data, e.g. rate of population decline, are not routinely used.

Red Data criteria have been revised and updated in 1994 partly in response to these concerns.

The causes of extinction

We are all versed in the many causes of extinction, whether the destruction of the Amazon rainforests to open up farmland, the hunting of tigers so that their bones can be ground into cold cures, or beach front discotheques disorientating hatchling turtles. The individual cases provide such a litany of woe that generally applicable patterns and processes are difficult to decipher. Underlying questions go unanswered. If rainforest is cut down it is seldom completely concreted over so why do remaining fragments and secondary forest suffer extinctions? Why are tigers so vulnerable to hunting when other animals are so difficult to eradicate? Why are turtles threatened by tourism but elephants benefit? The causes of extinction are the meeting place of the natural and human world.

The ecological causes of extinctions are natural processes that put wildlife at risk. These include ultimate causes that drive populations into decline or keep them rare, and the proximate causes plus those final straws that snuff out the last of a line. The pressures of human activities are a direct, proximate threat, e.g. exploitation, extermination, habitat destruction, introduced species, pollution and ecosystem breakdown. These threats are driven by four main characteristics of human systems: resource pressures, cultural attitudes, institutional policies and the failure of economics to value biodiversity properly.

Ecological causes of extinction

Natural processes can put species at risk from extinction. These ecological processes are divided into two: **ultimate causes** that drive decline and rarity and **proximate causes** that wipe out remnant populations.

Ultimate factors

Population abundance and geographic range

As a general rule species with a wide range have higher populations. Even if species are patchily distributed, species with wider ranges have higher populations within the patches they inhabit than more restricted species. This results in an ominous double jeopardy. Lower a population and its range will contract. Diminish the range and the population will fall. Why abundance and range are correlated is unclear. An intuitively attractive possibility is that species with broad niches can be widespread and locally abundant, able to exploit diverse resources. However, Gaston (1994) showed that niche breadth and abundance are not clearly correlated. Instead, resource availability may be a better predictor.

Abundance and range can also be analysed using **metapopulation** models. A metapopulation is a set of separate, patchily distributed populations linked by emigration and immigration. Models of metapopulation dynamics support the observation that wider ranges (more patches occupied) and higher populations (within patches) are linked, because of **rescue effects** (arrivals top up populations in patches near extinction) and higher populations compensate for losses during migration.

Worryingly there is a marked trend to smaller ranges and population sizes towards the tropics. Tropical species, the heart of Earth's biodiversity, may be naturally more vulnerable to extinction.

Patchy distributions within overall range

Populations can be unevenly distributed across the total range. Sometimes the differences are so extreme that high density areas act as **sources** topping up numbers in less favourable (**sink**) regions.

A population in decline, contracting in range, may fragment. Isolated, smaller populations are more vulnerable to extinction, perhaps because they are prone to local catastrophes and too far for rescue immigration or recolonisation. Fragments left behind in suboptimal, peripheral habitats, beyond immigration range from core sources, may be doomed even without additional human pressures. Alternatively some rare taxa may only survive in marginal sites since core areas have been lost. Either grim alternative is a problem for conservation, especially any attempt to reintroduce populations which will inevitably be doomed.

One large population is safer if demographic stochasticity is the main proximate threat to survival. Many, albeit smaller, populations are safer from the danger of large catastrophic disasters, e.g. fire and disease. Some patches may escape. This security decreases if events in individual patches are spatially synchronous. As with others of these ultimate causes evidence is sparse and contradictory. In general patchy populations show synchrony even if patches are hundreds of kilometres apart, so protection by this means may be rare. Given that larger populations are also less vulnerable to genetic failure and social breakdown what little evidence there is suggests that a few large populations are better than many smaller ones. (See Plate 16.)

Body size and trophic position

Evidence from animals suggests that big is bad. The megafauna overkills are an extreme example with human attention focused on large species that provided more meat, better trophies and that were a direct threat to people. Larger animals are often higher in food chains, so are vulnerable to any disruption to trophic levels lower down, whether losses of lower levels or damaging factors concentrating in higher levels such as the famous pesticide impacts.

Both these apparently simple inferences are very unreliable. Body size alone is a poor predictor of immediate vulnerability to extinction. Instead body, population and range size variously combine to give different patterns, varying even more with taxa.

Plate 16 *Habitat fragmentation. This South American tropical forest is a 1 kilometre sq. remnant, the last home of the three toed jacamar. The fragment may be too small to support a viable population, so the individuals are living dead, the species doomed once the survivors die. Hence the presence of the desperate birdwatchers*

Smaller species have contradictory attributes. Many show more frequent and larger population fluctuations (an increased risk factor) but tend to breed more rapidly and abundantly (decreasing risk). Part of the confusion arises because body size correlates with so many other ecological attributes of a species.

Colonisation ability
Colonisation, a process combining both dispersal to new sites and successful establishment, limits the ranges of many species both common and rare. Colonisation ability becomes an important factor if rare species are generally less effective colonists than the common ones.

Different species with good powers of dispersal have ranges of many different sizes. Poor dispersers show less variation, possessing small to medium range sizes. Given the links between range size and abundance poor dispersers may be at increased risk. Conversely some taxa have lost their powers of dispersal where emigration is almost inevitably a doomed enterprise.

Rarity shows a weak link with establishment ability but evidence is confounded by other constraints, in particular those that restrict the ability for rapid reproduction, which is generally useful for establishment.

Historical echoes
Species may be rare due to historical events of which we have little idea. In addition some lineages are more or less susceptible to extinctions though why is unclear. The vulnerable lines are those with small ranges and low abundance.

In conclusion the ultimate ecological causes of rarity appear idiosyncratic, with cause and effect difficult to distinguish and patterns and process varying across time and spatial scales. Combining the lessons of historic extinctions and recent losses, Raup picks out five generalities.

1 Species with small populations are more vulnerable.

2 Species with limited ranges are more vulnerable.

3 Extinction of widespread species is increased following an environmental first strike.

4 Extinction of widespread species is favoured by stresses outside their normal range of experience. The return time scale of the threat is important.

5 Simultaneous extinction of many species requires stresses that cut across ecological lines.

Proximate factors

Once a population is small or fragmented extinction can be caused by many processes.

- demographic stochasticity;
- environmental stochasticity;
- genetic failure;
- natural catastrophes.

Demographic stochasticity

Patterns of reproduction and survival are the demography of a population. Small populations (tens to hundreds of individuals) are vulnerable to random (stochastic) misfortunes, e.g. unable to find a mate, losses to a predator. Such events would be insignificant to the overall survival of a larger population but may finish off small numbers. The helmeted honeyeater, *Lichenostomus melanops cassidix*, a rare bird of Victoria State, Australia, is threatened by such accidents. The honeyeater is endemic to riverside forest, its range contracting in the face of human habitat clearance. The last ten colonies have collapsed down to just one of 60 birds. There is now an appreciable risk of extinction due to natural vagaries of mating success and nesting. In particular loss of a male during breeding will result in death of the pair's brood even if the female is still alive. (See Figure 4.6.)

Figure 4.6 *Distribution of the helmeted honeyeater in Australia. Only one colony (solid dot) remains. Open circles indicate old sites, now extinct. The bird is vulnerable to extinction from simple accidents, with all its eggs in one site basket*

Source: Redrawn from McCarthy *et al.* 1994.

Environmental stochasticity

Random fluctuations to a species' habitat, e.g. variations in weather or food supply, also represent a threat to small populations.

Again such changes would be little threat to the common or widespread which absorb local losses but isolated pockets of rare taxa can be lost with no opportunity to recolonise.

Research in the UK into the survival of pond fauna has shown that occasional drying out of a pond causes marked losses of permanent water species. Temporary pond species may colonise, their diversity often lower. However, a survey of temporary and permanent ponds in Oxfordshire revealed when ponds were ranked by rarity of species rather than diversity, four of the top five ponds, out of 39 surveyed, were temporary. One of the temporary ponds was home to water beetle, *Haliplus furcata*, classified in UK lists as in the most critical Red Data Book category 'RDB1, endangered'.

Genetic failure

Small populations risk genetic failure. Inbreeding loses genetic variability. Offspring become increasingly **homozygous** (i.e. similar genetic makeup), lowering short-term fitness such as resistance to disease and longer term genetic variation necessary to survive evolutionary pressures. This inbreeding may result from new, usually harmful, mutations or established disadvantageous characteristics finding expression as more individuals carry them. The process can snowball as more parents carry disadvantageous genes, lowering the chance that one parent may provide a dominant compensating gene to offspring.

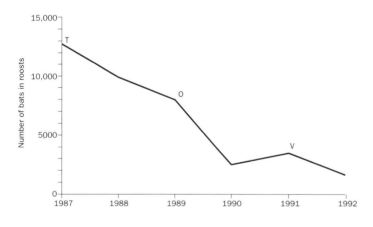

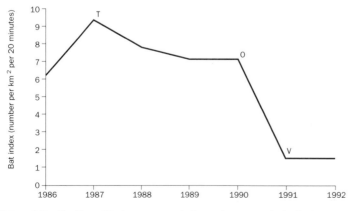

A study of 40 adders, *Vipera beras*, a fragment population in southern Sweden, isolated in a 1,000 by 20–500 m grassland some 20 km from the nearest populations, is an unusual example of evidence of inbreeding in the wild. Over the seven-year study period only 1 to 15 males and 2 to 18 females bred in any one year. Smaller litter sizes, increased deformity and stillbirth and marked

Figure 4.7 *Decline of bats, measured at roosts or counted when active, in American Samoa. T, O and V indicate different hurricane events. Bat abundance declined severely due to natural disasters*

Source: Redrawn from Craig *et al.* 1994.

genetic uniformity were characteristic of the population compared to adders from extensive populations. This degradation had occurred within the space of ten years since the population was cut off.

Natural catastrophes

Natural catastrophes could be regarded as extreme examples from a continuum of increasingly severe environmental stochasticity. But they do represent a qualitative step up, including the likes of volcanic eruptions, fire and flood. Volcanoes are explosively destructive. Fires are not simply hot but burn. Floods are a physically destructive force.

Natural disasters have been a factor in the decline of the Samoan and Tongan fruit bats (*Pteropus samoensis* and *P. tonganus*) in American Samoa. Numbers in the mid-1980s were estimated at 1,500 and 12,000 respectively. Hurricanes in 1990 and 1991 caused 80 to 90 per cent mortality, killing bats and wrecking forest habitat. Losses were worsened as human hunters took an added toll. The bats had always been harvested in small numbers but the hurricanes increased accessibility of roosts and many bats moved into village plantations. The post-hurricane harvest of 3,400 actually exceeded first estimates of the entire surviving bat populations. Bat numbers are now about 400 and 2500 respectively. (See Figure 4.7.)

Extinction. Human pressures and economic failures

The threat to biodiversity today, perhaps equivalent to a sixth major mass extinction, is caused by the impact of humans. Our actions can cause the decline and extinction of species in very direct ways, the **proximate causes** of extinction. However, these individual cases are driven by the four underlying **ultimate causes** of resource use, cultural attitudes, institutional failure and perhaps most pressing of all the failure of existing economic systems to capture the value of biodiversity fully.

Human impacts on biodiversity: proximate causes

Such impacts include:

- exploitation;
- habitat destruction;
- introductions and exterminations;
- pollution;
- ecosystem cascades;
- mismanagement and confusion.

Exploitation

Exploitation is the direct use of a species that drives down numbers. The famous examples, e.g. poaching of rhinos for their horns, sometimes eclipse the massive global trade in many species. Recent examples include the expanding Asian trade in terrapin and tortoise products for medicine and diet supplements. Trade has been cited as a major cause of the catastrophic decline of species in southern China and the Chinese markets have now expanded to exploit stocks from other Asian countries. Exploitation of at least six species is reported at unsustainable levels including trade in some species that is illegal under the CITES treaty.

Habitat destruction

Along with exploitation habitat destruction is the most widespread cause of direct impacts. Habitat destruction is typically the result of exploitation of habitat, so allied to the losses from the direct use of a species.

A typical case causing problems is the destruction of parts of the Khao Sam Roi Yot National Park in Thailand, a globally important area of coastal wetland, including freshwater marshes used by migratory birds. The area has been used historically by local people harvesting wild prawn. Government and industry have recently promoted intensive prawn farms which rely on artificially managed lagoons. Land prices have risen with farmers from outside claiming ownership by using dubious documentation and vital hydrology has been disrupted by artificial waterways. The building of hotels and golf courses has eaten up additional habitat. The National Park, which was set up in 1966, is vulnerable due to confusion arising from a 1982 review that actually expanded its size but created ambiguous boundaries. This destruction of primary habitat, combining conversion to farmland, coercive and aggressive acquisition and weak protection, is typical of many developing countries.

Introductions and exterminations

Humans have caused losses by the intentional extermination of species and the introduction of others. In 1995 African wild dogs in Namibia, numbered at 250 out of a total population in Africa of 5,000, were hunted by farmers following a pack kill of a calf and 32 dogs were killed. Wild dogs were still classified as vermin in Namibia and could be legally killed. By contrast the wild dog is one of only five mammals listed as Specially Protected Animals in Zimbabwe's Parks and Wildlife Act. Target species fare no better in the developed world. Remnant wolf and bear populations throughout Europe are still subject to persecution, even where officially protected.

Extinctions caused by introduced species have famously devastated many oceanic island systems. A more insidious problem, extinction through genetic dilution and blurring with introductions, has come to the fore in recent years. The world's most endangered dog, the Ethiopian wolf (*Canis simensis*), is vulnerable to this fate. Reduced to less than 500 individuals by disease such as rabies and canine distemper, both perhaps spread by domestic dogs, hybridisation with dogs is further eroding the species.

Pollution

Pollution may be one of the earliest forms of damage which has raised public awareness of environmental degradation but it remains just as widespread. Much of the South American rainforest is currently threatened by pollution of waterways from toxic waste products of gold mining, notably mercury and cyanide. The heavy metal mercury is used in small-scale extraction of gold and the waste is routinely dumped into rivers. Brazilian and Peruvian rivers have suffered. Guyana experienced one of the worst mining disasters in history in 1995, all the nastier given its predictability. Cyanide-rich slurry, including arsenic and copper, breached slurry tanks and spilled 3 million m^3 into the River Essequibo. Guyana's president declared an environmental disaster zone. He had previously welcomed the large-scale mining as turning 'our mudland into the gold land of the future'. Independent reports had warned of the likelihood of such breaches. Goldstar Resources Limited, the American mining company involved, explained their operations in Guyana more honestly. Goldstar had 'looked specifically at the Guyana Shield because of increased pressure from environmentalists and government in the USA'.

Ecosystem cascades

Human activities that, even if not directly destructive, spark effects which ripple through ecosystems as natural interactions are disrupted have become a common threat, especially of marine systems. Similar collapses have plagued the North Sea and north east Atlantic causing deaths and breeding failure of sea birds.

Another common cascade impact is due to overenrichment of nutrients in seas and lakes. The abundant supply can cause a switch from a diverse community to one dominated by a few species able to thrive on the enrichment. The destructive imbalance snowballs as more and more diversity is lost. This overenrichment is called eutrophication and has been linked to red tides, massive blooms of algae. Bloom species are often species that release toxic chemicals or cause further destruction when their death and decay deoxygenates the water. Recent deaths of the American manatee around Florida (over 150 out of populations of 3,000) have been linked to a toxic red tide.

Mismanagement and confusion

Recent improvements in conservation policy and practice can be set back, reducing hoped for protection. In 1996 the World Bank, Reconstruction and Development Division, pursued plans for a £1 billion dam, Nam Theun II, in Laos. The Global Environmental Facility (GEF), also managed by the World Bank was working on a £9 million conservation package to protect biodiversity hot spots in Laos that include endemics, e.g. the large antlered muntjac, only discovered in 1993, and a related deer which was found so recently that it still lacks a name. The GEF reported that the dam would cause significant environmental degradation and dropped funding, unwilling to participate in mitigation work while the main damage was still allowed to proceed. Work towards the dam continues, including logging and road building, despite evidence of flawed plans based on incorrect estimates of dam size and rainfall.

Human impacts on biodiversity: the underlying human pressures on biodiversity

The damage wrought to nature by humans is seldom driven by destructive spite. The particular examples of individual threat outlined above are not arbitrary or accidental. Rainforest is not cut down simply to wipe out unknown canopy beetles. Introduced species are not unleashed on oceanic islands specifically to kill strange flightless birds. These activities are the result of deeper forces and failures. To understand biodiversity loss it is vital to understand the economic and social pressures.

Four main pressures and problems drive human destruction of the environment: resource pressures, cultural problems, institutional failures and economic failures.

Resource pressures

These resource pressures include human population growth and the drive to globalisation.

Human population growth

The global human population continues to grow rapidly. Estimated at over 5.6 billion in the mid-1990s, the total is reckoned to double by 2052 or, assuming low fertility rates, to increase by 30–40 per cent. The developing world's youthful population will maintain the momentum, adding a billion each decade. Increased population has a direct impact by consuming more resources unsustainably and limiting regeneration. Indirect effects include poverty, migration (especially into undeveloped wildlands) and social breakdown. Population growth causes habitat loss and uses up ever more natural production. Humans consume 40 per cent of terrestrial primary productivity.

Consumption is growing faster than population. Examples of increased consumption between 1950 and 1990 include fish catches up by 44 per cent, fertiliser use up 970 per cent and natural gas production up 1150 per cent. In the meantime more of the world's populations become consumers. Increasing urbanisation degrades biodiversity by habitat loss, fragmentation, introduction of exotic species, pollution, drain of resources, disruption of natural geochemical cycles and conversion of adjacent land for farming or suburbs as some try to escape the crowded cities. Of the world's population 45 per cent lives in cities. The total of 190 cities with a population of 1 million plus in 1975 is predicted to reach 400 by the year 2000. (See Box 16.)

Population movements in response to economic need add to the damage. Cultural and institutional restraints collapse among mobile communities as they move into undeveloped land. Settlers often lack any sense of a link to the land which fosters a sense of responsibility for resources. The steady erosion of the Amazon wilderness is a grim example.

Drive to globalisation

The diversity of crops, agricultural techniques and production systems around the world have been replaced by an increasing reliance on a small number of crop

---

Box 16

Human Impacts

Resource pressures: bushmeat in Equatorial Guinea

Forest animals hunted for bushmeat are a widely used resource in West and Central Africa. Subsistence hunting is increasingly replaced by commercial hunts to supply towns. Recent studies of two town markets in Equatorial Guinea, including on Bioka Island a biodiversity hot spot supporting endangered and endemic subspecies primates, have revealed the sheer numbers of animals killed. Over 18,000 carcasses were brought in for sale over ten months. The tally concentrated on a few species, 9 antelope, 16 primates and 3 rodents, with evidence of preference for size, neither too small or large. Local taboos against eating some species, e.g. Colobus monkey, were breaking down as hunters ranged further afield into new areas. Hunting had now spread into reserve areas as improved guns and transport increase efficiency and quarry decline near villages.

---

species. These can be traded in a global market. The globalisation of economies has broken the links between the management and consumption of resources. Agriculture has undergone specialisation with an increasing dependency on a few crop species. Many are now effectively biosphere people and able to escape problems of overexploitation by shifting to a new market for resources. Individual countries that adjust economic policies to penalise destructive exploitative trade by their own industries open themselves up to undercutting by competition from countries that do not. Free trade and export driven economies undermine local markets. This is exacerbated where countries consciously court foreign investment, perhaps requiring disruption of local social and cultural traditions to succeed. Global trade can be a direct threat to valuable species. Illegal trade is still rife. When a single Russian Siberian tiger can be worth £40,000, with the poacher paid £660, enough to support a family for two years, and its bones, organs and skins traded throughout Asia, then the global market is an irresistible threat.

Cultural problems
These problems include inequality of ownership and property rights and cultural attitudes.

Inequality of ownership and property rights
Iniquitous ownership, control and trade work favour of degradation and extinction. Globally and locally a minority owns most resources and reaps the benefits of exploitation while many bear the costs.

Equality depends on property rights. Biodiversity is a resource and property rights are the ability to secure, use and derive value from the resource. There are four main types of ownership rights: **open**, unregulated; **common**, regulated by rights and responsibilities; **private**; and **state**. These systems mesh with legal systems that can be either **custom and tradition** or **formal legislation**. The third factor that influences equity is **social position**. For example, many clashes in the developing world pit indigenous smallholders against outsiders and large companies. No one property system has a monopoly of virtue. The important factor is security of tenure, bringing a sense of responsibility to future generations. This permits a long-term view rather than short-term cut and run. The rights of future generations, inherent in many definitions of sustainability, have proved especially difficult to build into economic processes.

Overconsumption by the developed world is a driving force for losses as the developing world struggles to exploit resources, often destructively, for a short-term gain. Meanwhile the developed world is able to conserve resources. At its crudest it may pay developing countries to cut down forests and put the money into capital projects or even the bank.

Cultural attitudes

The diversity of attitudes to biodiversity is astonishing but three broad cultural patterns exist. Many traditional societies combine use with a sense of nature's value and importance to sustain their society. There is utilisation and respect, with taboos and beliefs effectively creating rules which conserve the natural system. Their attitudes and behaviours reflect awareness that the resource base is limited, that substitutes are not readily available but that they do have control over their catchment thus being in a position to manage wisely. Such cultures have been termed **ecosystem people**. Conversely, **biosphere people**, typically in the developed world and urban elites of developing countries, treat biodiversity as a resource which can be exploited to destruction. Resources are drawn from a global catchment. New resources can be opened up once others are exhausted. Biosphere cultures may lose the sense of rights of future generations in the expectation that new resources will be found and new technologies developed. The ecosystem to biosphere culture shift spawns a third category, **ecological refugees**. These are ecosystem people who are forced to move to new sites. Shanty towns of many developing world cities are the obvious example, but historic ecological refugees include the early European colonists of North America and settlers on the western frontier of the USA, plus colonists of many oceanic islands. (See Box 17.) Ecological refugees will exploit a limited resource base but have lost any integration with their locality. They may simply lack any knowledge, as well as having little real control or sense of security into the future. This is a recipe for degrading biodiversity. (See Plate 17.)

Institutional failure

Institutional failure comes about through institutional weakness and lack of knowledge.

Institutional weakness

All the good intentions, treaties and laws in the world are ineffective if the will and means to conserve biodiversity are lacking. Many developing countries with fragile political structures that are open to coercion by force or corruption are very vulnerable. However, even well-established schemes that are apparently secure can

Box 17

Human impacts. Economic and social forces; the passenger pigeon, Ectopistes migratorius, and the American buffalo, Bos bison athabascae

The extinction of the passenger pigeon and the near extermination of the American buffalo are apocryphal examples of species loss. Both cases show the interplay of economic and social factors.

Culture Both species occurred in proverbially vast numbers on the frontier of the expanding USA, an area populated by the increasing settler population, ecological refugees and native Indian cultures, ecosystem people.

Inequality and property rights The pigeon and buffalo were an open access resource for the settlers, while inequality of power undermined the property rights of the native Indians. Buffalo hunting was in part an attempt to undermine the native Indian cultures, examples of biodiversity culture, reliant on the buffalo as a resource and spiritual focus.

Economic forces Commercial hunting of pigeon and buffalo went on until the 1880s, the last large-scale buffalo hunt took place in 1884. Economic forces sustained the hunts even as numbers declined. The expansion of the railroad network improved access to distant stocks and movement of carcasses and products, so that demand could increase while prices remained low. Buffalo hunting was encouraged by failures of other leather supplies and improved tanning technology. The market was also easy to enter and leave, allowing hunters to switch in and out depending on availability of other jobs. Ultimately conventional economics suggests that extinction should still be avoided as populations become so low that the costs of hunting become too great relative to value. In neither case did this happen. There is no evidence from prices which remained broadly stable even in the last year of hunting. Evidence of local employment also suggests buffalo hunters did not anticipate any final collapse until it had happened. There were professional pigeon hunters even after wild stocks were gone. The buffalo was saved by the establishment of 18 herds, most fewer than 10 animals, between 1873 and 1919. The passenger pigeon lingered on in at least three captive zoo stocks, including more than 20 in Cincinnati Zoo where the last one died.

be undermined. A recent example has been the fate of Project Tiger in India. Set up in 1972 Project Tiger co-ordinated a system of reserves focused around the tiger, *Panthera tigris*, which also benefited many other species. Initially successful estimates of Bengal tiger numbers rose from 1,800 to 4,300 in 1992. Political upheaval, particularly the deaths of Prime Ministers Indira and Rajiv Ghandi, started a rapid collapse. In 1993 numbers fell to 3,750 as political will and patronage decayed. Habitat was also threatened by political policies to open up the Indian economy to foreign companies. Funds dried up and poaching began, even in famous tiger reserves. When caught the dealers in tiger products repeatedly evaded imprisonment. By 1995 only 2,500 tigers were left, with estimates of one tiger killed each day. A survey of Project Tiger schemes revealed that 80 per cent now lacked any armed anti-poacher patrols and 75 per cent did not receive funds on time. In October 1995 a meeting in Delhi set out to revive the work: 'Save the Tiger, save the jungle, save India'. Corruption, greed, intimidation and indifference are powerful enemies.

Plate 17 *Grey seals (*Halichoerus grypus*) on the Farne Islands, north east England. The UK supports internationally important breeding stocks of the Grey Seal, but their conservation has become politically sensitive due to the belief of some fishermen that the seals take too many of the dwindling North Sea fish stocks*

Lack of knowledge

Our incomplete inventory of life inevitably endangers biodiversity. What expertise exists is unevenly spread, concentrated in the developed world. Lack of knowledge can even threaten well-known species of the developed world.

The extinction of the British race of the large blue butterfly, *Maculinea arion*, is a classic case. First recognised in the eighteenth century in colonies scattered throughout southern England the Large Blue was showing local extinctions and range contraction by the 1900s. The species became the target of intense conservation work. Nonetheless the last colony was lost in 1979. Over-collection was blamed but evidence was weak. Half the known sites had been lost to direct habitat destruction but many other protected colonies had also been lost. Too late we had understood the butterflies' requirements. The caterpillars mature as parasites inside ants' nests. Several species of ant will pick up young caterpillars and carry them back to a nest. This life history was only worked out in 1915. In the 1970s it became clear that one ant species, *Myrmica subuleti*, is a much better host from the caterpillar's point of view. Other ant hosts attack the parasitic caterpillars more readily. *Myrmica subuleti* thrive in swards with some grazing. Without grazing the tall vegetation cools the microclimate so other ant species dominate. Intense grazing reduces the grass sward so yet other species move in. By this time the last few large blue sites had had grazing halted as a conservation measure or, where rabbits were present, myxomatosis reduced their impact. Our ignorance had doomed the large blue. In 1983 European stock was reintroduced but the unique British race was no more.

Economic failures
Economic failures include market failure and interventions and subsidies.

Market failure to capture full value of biodiversity
This problem afflicts both the most affluent northern hemisphere and the poorest developing economies. The mechanisms are similar even if the detail of habitats and species varies around the world. It is also perhaps a less obvious threat than some of the glaring problems of local politics or agricultural change. But though less tangible it is all the more dangerous and to some extent drives the other problems.

Markets do not capture the full value of biodiversity, its value to society globally or locally. Generally an individual or an organisation will gain financially from exploiting resources now, given the uncertainty of the future. Natural capital, including biodiversity, is turned into manmade capital (generally financial) or owned. Not only does the individual gain the immediate profit but very often many of the costs (e.g. effluent) can be dispersed into the environment. The full cost of exploiting the resource (e.g. effluent treatment or containment) is not carried by the individual beneficiary. A problem shared is a profit increased. In the meantime society at large ends up paying for the burden of any such costs, perhaps by direct societywide payments (e.g. taxes to support anti-pollution management and treatment), but also more insidiously through loss of environmental quality and ecosystem function. Such damage can accumulate and its effects only become apparent when damage has been done. The so-called free market is even freer than supposed since individuals can dump some of the costs of their activities onto everyone else. Such gratis get-outs are called **externalities**, because the costs are borne by others. The benefits of intact healthy biodiversity have no market. They are missing from economics.

The individual also loses out, suffering the same loss of environmental quality as everyone else, though financial security may buy some protection from environmental degradation. The short-term gain may be incentive enough to compensate. Individual versus global costs and benefits from exploitation diverge so biodiversity is lost. The problem may be exacerbated where the general benefits give little local return or even inflict costs. The former is exemplified by demands for conservation of the tropical rainforests, largely found in developing countries, to maintain atmospheric quality in the industrialised developed world. (See Box 18.)

Intervention and subsidies
Institutional policy can drive the loss of biodiversity. Interventions are often seen as politically easier than developing economics that fully value biodiversity, but even well-intentioned interference can be damaging.

Box 18

Human impacts and economics

Valuing wildlife: game versus cattle ranching in Zimbabwe

A fine example of economic failure to value biodiversity is the contrast between cattle and game ranching in arid regions of Zimbabwe. Cattle stocking rates are based on a very short-term view of income and do not account for the long-term decline due to chronic overstocking. This is found in 70 per cent of cattle only farms and degrades the habitat. Studies of cattle versus game ranches in the semi-arid midlands compared economic projections taking into account over-stocking impacts over the medium term. Economic models that did not account for overstocking suggested cattle were almost always more profitable. When degradation from overstocking was included wildlife ranching profits were hardly reduced. In effect game was hardly ever over-stocked and even where it was it did little damage. However, overstocked cattle farms showed marked decline in profits which, projected into the future, suggested economic failure.

Direct subsidy of destructive activity occurs worldwide. Not only does biodiversity suffer from undervaluation but such policies heap on additional bias in favour of exploitation rather than conservation. The environment is degraded and sustainable economics is undermined. The global total for such destructive subsidies has been estimated at $600 billion per year. Examples include subsidy of intensive agriculture in Europe that encourages habitat destruction; subsidy of conifer afforestation using non-native species on bogs in Scotland during the 1980s; subsidy of forest clearance in Brazil for livestock. Subsidies continued to rise through the 1980s. Proportions of farmers' incomes made up from subsidy in different continents between 1981–4 versus 1989–92 included Australia 11 versus 12 per cent; European Community 32 versus 46 per cent; and the USA 27 versus 27 per cent. Subsidies are rife in the developing world, often in support of damaging practices, e.g. percentage of pesticide price subsidised in 1993: Senegal 89 per cent, Ecuador 41 per cent, Indonesia 82 per cent. Resources used for subsidy are also diverted from more sustainable uses such as education and research, thus dooming deprived populations to exploit more land using unsustainable techniques.

Since the early 1990s increased awareness of damage done by direct subsidy of destructive activities has resulted in a change in policy, e.g. tax subsidy of afforestation of Scottish bogs and clearance of Brazilian forests for cattle ranching have both been revised.

Valuing biodiversity: economics for conservation

Methods for costing value of biodiversity

The true economic value of biodiversity is a new field in which some possibilities are emerging. If the true value of biodiversity can be properly realised and included in economic mechanisms then a very powerful tool for conservation is available. First it is necessary to distinguish the types of value which biodiversity may represent.

1 **Use value**
   - direct use, e.g. harvesting, subsistence, tourism;
   - indirect use, ecosystem function;
   - future use, potential uses, insurance against the unknown.

2 **Non-use value** (the benefits of not using biodiversity). Essentially this is the amount we are willing to pay to conserve, willing to accept in compensation for not exploiting or willing to forgo by not maximising a financial return, to keep the biodiversity. Such concepts include:
   - **bequest value**, the classic 'our children's children' argument;
   - **existence value**, willingness to pay to conserve a species which we may not even see, content to know that it survives;
   - **option value**, potential uses, direct, indirect and insurance against the unknown.

Several approaches have been used to put financial values to these concepts.

Changes in productivity and economic gain

These are direct attempts to value economic returns associated with sustainable management compared to degradative exploitation.

Contingent valuation

Contingent valuation is the establishment of a financial value based on willingness to pay, willingness to accept or willingness to forgo. Values based on the willingness to accept a reduced return often exceed those based on willingness to pay to conserve. Willingness to accept is an abstract loss of future returns which are not in the bank, while willingness to pay involves handing over what has already been earned. The term contingent reflects the severe biases that can ruin such valuations, commonly based on surveys and questionnaires.

Hedonic pricing

Hedonic pricing relies on the expertise of economists to tease out the value of natural phenomena and environmental quality from within the total price of a resource. For example, a house in beautiful unspoilt countryside in the UK would fetch a different price from exactly the same house in the middle of a grim urban conurbation. If this separation can be achieved general models of demand and value can be built. This is tricky and requires skill and sufficient data on prices. The technique only works for products that are being marketed so that an overall price exists which can be broken into components.

Travel costs

The value of biodiversity as a recreational resource can be estimated from the costs that tourists are willing to pay to visit sites and view wildlife. Some costs may be easy to specify, e.g. air fare, hotel, park entrance fee, but holidays are often not solely to see wildlife. Some costs can be difficult to specify. In addition the damage done by tourists, whether through resources used or uncosted externalities, should be accounted for. In some cases the income can be expressed directly as a financial value for the land area involved, allowing comparison with potential alternative land uses.

Valuation of substitutes

This is a technique used to value ecosystem functions that relies on costing the technology which would be needed to replace degraded biodiversity. One role of wetlands is as a sink for excess nutrient run-off, protecting rivers from such degrading effluent. The cost of drainage infrastructure and machinery necessary to achieve a similar diminution in nutrient pollution can be calculated directly.

The economics of conservation

The last decade has seen an explosion of work which tries to tie in the intuitive feeling that biodiversity is valuable with financial measures of this worth. Although techniques are tentative, produce differing estimates and depend on often patchy data, the economic arguments for conservation have shifted from hazy notions to startling estimates of the financial value of biodiversity.

Genetic resources

The Green Revolution of the 1960s and 1970s raised productivity of wheat and rice by some 60 per cent. This was in part due to genetic improvements to cultivars. As production rose so did fluctuations in yields, in part due to the narrow genetic base of crops used. In the 1970s four varieties made up 75 per cent of the USA potato crop, an echo of the Irish Potato famine of 1846, when reliance on one variety that was devastated by potato blight led to 1 million deaths and emigration of a further 1.5 million. Genetic uniformity has resulted in crop vulnerability throughout recent decades, e.g. US 1950s wheat rust, 1970s maize corn leaf Blight, 1984 Florida citrus bacteria infections.

Genetic diversity in crops is an insurance against the vulnerability of monocultures and is vital since 90 per cent of the world's food base consists of four species of plant and five of animals. Genetic variety may also reduce the need for chemical supplements, cutting costs and environmental damage. The genetic diversity is also the raw material for manipulations. Crop varieties and livestock breeds can be improved for particular conditions, for resistance to pests and disease, to boost productivity and product quality and to aid cultivation techniques. The US tomato industry received a multimillion dollar benefit by inclusion of a jointless stalk gene found in the wild only on the Galapagos Islands. Estimates of the value to agriculture can be made in several ways. An example of comparison of costs for research versus resulting gain come from corn breeding in the USA. In 1984 $100 million was spent on research but $190 million was earned as a result. Improved crop yields are an obvious measure. In the 1980s US total crop productivity rose due to new cultivars by $1 billion. In Asia rice and wheat productivity in the Green Revolution rose by $1.5–2.0 billion.

Genetic diversity is highly valued for pharmaceutical products. Plant products can be used directly or as raw material to refine into therapeutic derivatives or as the inspiration for synthesis of artificial analogues. In the latter case the value of biodiversity is as pure information. Estimates of the value of plant-based pharmaceuticals include some extensive work in the USA where plant products were used in 255 of drugs prescribed between 1953 and 1973. The plant components were estimated as worth $1.6 billion, rising to $9.8 billion in 1980, then $18 billion in 1985. Since the bulk of products came from only 40 plant taxa the 1953–73 estimate is equivalent to broadly $200 million per species. The value can be doubled when hospital use is included. Add to this the savings from illness avoided or cured and therefore work days not lost and estimates of the annual benefit in the mid-1980s were between $34–300 billion. These data have been taken further. Estimates suggest 5,000 species of plant had to be examined to find these 40 most useful (i.e. 1 in 125). Assuming the same proportion holds globally (we are out on the thin ice of assumptions here) for every 1,000 species of plant extinct eight could have yielded very useful drugs. The return to pharmaceutical prospectors from new drugs found in the Costa Rican forests have been cited as $4.8 million per drug. In this light the loss of economic options from plant extinctions is high and the retention of option values from conservation are valuable. Recent advances in **biotechnology** and **bioassay** have spurred the practical prospecting for new plant-based drugs, e.g. Taxol, an anti-cancer drug derived from bark of the Pacific yew, *Taxus breviflora*.

Taxonomic diversity

Many species have a financial value from direct use e.g. as a sport target or tourist attraction. Their value is often directly estimated from financial returns linked to sport or tourism. Species also have a vital but still poorly understood role in the function of ecosystems which is much trickier to estimate. Ecosystems may depend on a small set of **keystone** or **driver** species, the other taxa present **passengers**, an excess sometimes described as **ecological redundancy**. However, passenger taxa may become important if the environment changes. The link between species and critical ecosystems function is contentious but the functional diversity of species, what they do rather than taxonomic relatedness, may be important for ecosystem **resilience**. This resilience is the size and frequency of disturbance that an ecosystem can withstand without substantial change. Such roles are very difficult to cost.

A financial value can be put on direct use species. Estimates of the value of the African elephant in the late 1980s range between $22–30 million per annum, based on several different measures of costs borne by tourists. Similarly the value of Kenya's protected areas, their attraction largely due to species, has been estimated as $540 million in 1994, greater than economic returns from any alternative use. Where the intrinsic, non-consumption value is greater than that of any actual or potential destructive exploitation then economics becomes a very powerful incentive for conservation. But two key steps are required. First, the financial value must be made explicit. Second, the value must be realised by local people since they bear the costs and forgo the opportunities from not destructively exploiting the species.

Ecosystems

Like species, ecosystems can be valued both for direct gains and indirect benefits, particularly ecosystem functions. Central to all valuations is the comparison of benefit from conserving an ecosystem versus the costs of management and of opportunities lost.

The financial value of tourism has proved to be an effective approach. Global ecotourism in 1988 was estimated as worth $233 billion, of which half could be linked directly to biodiversity. Detailed studies of tourism to Costa Rican rainforests in the early 1990s suggested their value was $1,250 per hectare, against costs of conservation management of $30–100 ha and value of adjacent farm land of only $30–100 ha. Economically conservation would be the best option, so long as the benefit reached local people.

Valuing ecosystem functions has proven trickier. Functions comprise **regulation**, e.g. climate, watershed processes; **production**, e.g. food, fuel, oxygen; **carrier**, e.g. space for agriculture; and **information**, e.g. spiritual, scientific roles. The ecosystems functions create resources or carry out processes with substantial economic benefits. The benefit of Cameroon rainforests as regulations of watershed processes has been costed as equivalent to $54 hectare; Brazilian forests as carbon stores at $1,300 per hectare per year; restored Swedish wetlands as nitrogen sinks at $18.7 per kg of nitrogen fixed.

Most attention has focused on tropical forests and attempts made to measure the **total economic value** which must capture the direct, indirect, option and existence values. Even when this is not possible values can be put to different aspects of their benefit: value versus alternative uses; local use values and individual ecosystem functions.

Value versus alternative uses
Even if complex often intangible values are difficult to pin down comparisons to alternative uses can be revealing. A classic study of the Korup National Park, Cameroon, estimated benefits from use, function, trade and tourism of £23.6 million versus set-up costs and lost opportunity of £16 million, a net gain of £7.6 million. Similarly a study of Bacuit Bay in the Philippines compared benefits from multi-use

including logging to those without logging, tourism and fisheries on an adjacent coral reef. Note in this case the explicit benefit to another ecosystem, the reef. Depending on precise estimates of future returns the benefits of not logging over ten years were $11.5–17.5 million.

Local use values

Many studies from around the world have assessed the value of local sustainable use of forest. The range of values is considerable but many are high. South American examples (all given as value in $ per hectare per year) include an experimental Caiman harvest in Venezuela, 0.75, Peruvian villagers' multi-use of forest resources, 16–22, and wildlife value of Ecuador rainforest, 120. Indian examples include forests as a source for domestic elephants 3.0, village household multi-use 50, and India-wide gross benefits from local use of food and medicine plants, 117–144.

Individual ecosystem functions

Estimates for the value of specific functions have concentrated on the role of forests as a carbon store. At a global level one hectare of tropical forest has been valued as $200 and if cleared making contribution to global warming via carbon release of $2,000–4,000.

The Amazon as a whole, acting as a carbon store, has been valued at $46 billion, which when combined with a direct use value of $15 billion and existence value of $30 billion adds to a total value of $90 billion. At a local scale estimates of the value of rainforests (all as value per hectare per year), from different sources using different techniques, include Costa Rica, $102–214, Thailand $400, value to US of foreign rainforests, $500.

So long as economics can capture the total value of biodiversity in particular externalities and ecosystem functions and ensure that local people benefit, economics can be a very powerful incentive tool for conservation.

Local economic incentives and conservation

Some of the most effective progress linking biodiversity and economics has developed to rebuild links between local control and benefit and conservation of wildlife. This is best illustrated by two schemes, one a project founded by a voluntary conservation body focused on a special site in Cameroon, the second of Zimbabwe's national programme linking wildlife and local people, CAMPFIRE.

The Mount Kupe Forest Project, Cameroon

Mount Kupe is a granite peak swathed in rainforest, covering a mere 21 sq. km. Designated as the Mount Kupe National Park and famed for its bird life, it is a regional biodiversity hot spot focused on eastern Nigeria and western Cameroon. There are at least fourteen endemic species, most famously the Mount Kupe Bush

Shrike, *Malaconotus kupensis*, a Lazarus taxa, thought extinct until its rediscovery in 1989. In addition the grey-necked picathartes, described as a cross between a crow and vulture that skulks around in caves, is a special draw. There are at least 300 other bird species in the Park. The area was conserved in part by traditions of the local Bokossi people who revered the mountain as the source of wealth and well-being. Outsiders now outnumber indigenous locals and the Park is under pressure from clearance for banana plantations and hunting meat.

Plate 18 *Mount Kupe Forest Project, Cameroon. Publicising the benefits of conservation. A local building sports a picture of the Mount Kupe Bush Shrike, a flagship species for conservation of the site*

In 1991 BirdLife International, a conservation charity, started the Mount Kupe Forest Project (MKFP) following preliminary meetings with the local villagers. The project intends to conserve the montane habitats by facilitating the economic benefit to local people from the presence of intact forest. The scheme relies on income from visiting tourists who stay with families in the adjacent village of Nyasoso. (See Plate 18.)

The first tourists came in 1991–2. The 41 visitors stayed a total of 160 tourist nights. In 1992–93 numbers rose to 94, staying 440 tourist nights. The income to the village was estimated as £5,500 in total. In 1994 the cost of one night's stay per person was £6, with additional income to locals acting as guides. This money goes straight to the villagers. In addition tourists' donations to the project raised £323, of which 40 per cent was used on trail and new campsite management, with 60 per cent going to the village. Birdwatchers have proved to be the most

Plate 19 *Mount Kupe Forest Project. Visiting birdwatchers are asked to talk to local children about their enthusiasm for wildlife and its importance and to let them have a go with binoculars and telescopes*

valuable visitors, staying more nights to see more birds. To date only 21 visitors have seen the shrike, a notable addition to any birdwatcher's list. (See Plate 19.)

There are many additional benefits on top of the cash. An education programme includes awareness of conservation. Pride in the project is promoted by noticeboards displaying foreign publicity. A pen pals scheme, run by the Royal Society for the Protection of Birds, linked local children to schools in the UK. A travelling library, a rickety charabanc, circulates material to more schools and has become a much loved icon for the children. Visitors are encouraged to talk to children about the importance of the birds and let them try their binoculars and telescopes. The scheme also trains up teachers and produces a teaching magazine to broadcast impact. Adult education includes agricultural and health advice. Some locals have become very involved with the project, gaining experience from trips abroad and training in ecological monitoring so that skills are transferred into the village. The project even sponsors a football team.

The CAMPFIRE programme, Zimbabwe

One of the most famous national schemes to link conservation of biodiversity with economic benefit is the Communal Areas Management Programme For Indigenous Resources (CAMPFIRE).

Before colonisation the black population had utilised wildlife through systems of common ownership carrying rights and responsibilities. These links were broken in the colonial era and much of the local population was forced to live in Tribal Trust Lands (latterly named Communal Area) which were on economically marginal terrain. The CAMPFIRE scheme reinstates control by and benefits to local people, creating an incentive for conservation of biodiversity in their area rather than destructive exploitation.

CAMPFIRE's aims are: long-term development; management and sustainable utilisation of natural resources in Communal Areas; management of resources by placing responsibility and custody with local people; to allow communities to benefit directly from exploitation of resources in their Area; to provide administrative back-up and advice. Alternatively, and better capturing the spirit of the scheme, Zimbabweans talk of the choice between wildlife as relic or resource, of wildlife paying its way, 'those who pay the social cost of living with wildlife should reap the economic benefit'. The costs can be high with crops destroyed, livestock lost and people killed every year by dangerous game such as elephants. CAMPFIRE schemes vary in detail of wildlife involved and utilisation (meat, live capture for sale, big game hunting, photographic safari, fishing). They are developed through a protocol which links local people to the Department of National Parks and Wildlife Management (DNPWLM) working with District Councils. The Councils should ensure that benefits reach local people on whose land the wildlife thrive, that locals should have full control over use of the money and that Councils should be accountable for plans and revenue.

A typical early example is the Dande Communal Area CAMPFIRE, begun in 1987. Dande CA is in north-east Zimbabwe, wedged between a Safari Area and other Communal Areas. Soils are poor, rainfall is erratic and Tsetse fly occurs, transmitting sleeping sickness. Ironically Tsetse control in the 1980s saw a 30 per cent rise in local populations without any real improvement of the economic base. In 1987 the DNPWLM advised locals on options and opportunities and helped with initial administration and surveys of wildlife to draw up quotas. The Dande Council co-ordinated local Wards. The northern half of the Communal Area was leased to a commercial big game hunting safari and the southern half used to set up a hunt owned and managed locally, to boost local skills. Problems arose with initial investment for campsites and lack of marketing so grant aid was provided by the Zimbabwe Trust. In 1989 the first revenue rolled in. The local owned safari raised Z$299,000, but cost Z$232,000, so profit was Z$6,700. The private company proceeds added Z$168,000 and grant aid Z$98,000. The total profit of Z$333,000 was split among 7,000 people, about Z$47. This income is substantial. The statutory minimum wage in Zimbabwe was Z$240, but in rural areas local people earn much less. Income per household from other CAMPFIRE schemes at that time ranged between Z$3 to Z$2,306, the latter enough to buy 38 goats. (See Figure 4.8.)

There remain problems, often associated with distributing such largesse, but the major ones have been linked to international attitudes to conservation. The Dande CAMPFIRE scheme is not untypical in its reliance on big game hunting. In 1993 big game hunting in Zimbabwe was worth US$4 million, of which elephants made up 70 per cent. Killing elephants is politically fraught. The Zimbabweans and other southern African countries have been pushing to reopen the ivory trade. Opposition has been fierce. Legal ivory trade might open up poaching and laundering of ivory, thus wiping out remnant elephant populations elsewhere in Africa. Elephants have also gained status as charismatic sentient creatures, beloved of TV audiences to whom the very idea of shooting them is abhorrent. Advocates of the Zimbabwean approach have been accused of a mixture of naivety and self-serving careerism. Shooting wildlife to conserve it remains a battleground of competing attitudes.

Conservation's targets

Many Zimbabwean mammals have an explicit price on their heads, the minimum trophy fee a hunter must pay to shoot one is, for example, male elephant £1000, male buffalo £170, Giraffe £250 (1991 prices). The animals represent money on the hoof to local people, targets for hunters, the hunters themselves targets for specialist holiday companies. This is not the image that many people have of conservation. There are more familiar and generally less contentious approaches to conservation via treaties, reserves, reintroductions and biodiversity. However, given that the threats to biodiversity are so firmly rooted in human pressures and economic failures, the examples of Mount Kupe and Zimbabwe represent a lesson which we should not ignore. Conservation has undergone a sea change in recent years. It is no longer simply a matter of protecting animals on nature reserves but of conserving and utilising biodiversity in all its forms, as Chapter 5 will show.

Many people think of wildlife
as a pest or problem . . .

but . . . .

Hunting
Safari hunters pay large sums
of money to shoot wild animals.

Figure 4.8 *CAMPFIRE programme literature aimed at Zimbabwean villagers to promote
conservation of wildlife by a utilisation as a sustainable resource*

Source: Zimbabwe Trust.

Summary

● Current extinction rates are difficult to estimate but many techniques converge on estimates suggesting losses are much higher than natural background rates.

● Ecological patterns suggest that there are general processes which increase vulnerability to extinction and cause final extirpation.

● Accelerated losses are due to human pressures of population growth, culture, institutional and economic failure.

Discussion questions.

1 How many species of vertebrate have been driven to extinction in your country and with what consequences?

2 Devise an economic value for a common wild animal or plant in your region.

3 Is extinction an ethical problem?

See also

Historical extinctions, Chapter 1.
Ecological processes creating diversity, Chapter 2.
Economics and conservation, Chapter 5.

General further reading

Extinction Rates. J. H. Lawton and R.M. May. 1995. OUP, Oxford.
Detailed review of extinction rates and risks, past and present.

Rarity. K. J. Gaston. 1994. Chapman and Hall, London.
Fascinating review of rarity, definitions, patterns and processes.

The Economic Value of Biodiversity. D. Pearce and D. Moran. 1994. Earthscan, London.
Short but sweet introduction to possibilities and problems for economic valuation of biodiversity.

'Generation, maintenance and loss of biodiversity'. R. Barbault and S. D. Sastrapradja. 1995. In V. H. Heywood (ed.). *Global Biodiversity Assessment*. Section 4. CUP for UNEP, Cambridge.
Detailed review of the ecology of extinctions.

'Human influences on biodiversity'. J. A. McNeely, M. Gadgil, C. Levegue, C. Padoch and K. Redford. 1995. In V. H. Heywood (ed.). *Global Biodiversity Assessment*. Section 4. CUP, Cambridge.
Detailed review of causes and consequences of human pressures on biodiversity.

5 The conservation of biodiversity

Conservation means much more than guarding charismatic species inside fenced reserves. This chapter covers:

- New attitudes to conservation
- Legislation, treaties and funding
- Protected areas
- Biodiversity in captivity

Conservation as a recognisable coherent scientific movement coalesced from diverse piecemeal actions and insights after World War II. Attempts to secure the scientific understanding and management of species and habitats for conservation led to the foundation of international (e.g. WWF) and national (e.g. Nature Conservancy in the UK) expert organisations. Since the 1980s conservation has undergone a sea change, driven in part by the multifaceted nature of biodiversity. Conservation has become much more than the province of scientists. Conservation science and environmentalism swapped insights. In particular conservationists appreciated that the factors driving habitat loss and species extinctions lay outside the immediate horizons of biology. The ultimate human impacts were the product of economics, politics and society. Conservation expanded its horizons with the realisation that protection of biodiversity, whether individual genes, species or ecosystems, could not be achieved by science alone, however expert, however committed. Conservation could not work in isolation. Integration of conservation into the wider realms of economics and politics was vital. In an analysis of the recent progress of ecology and conservation Max Nicholson, the influential originator of much UK conservation, points out that the task involves the reconciliation of three apparently intractable elements: people, area and the biosphere. The science of biodiversity, the mysteries of ecology and evolution were easy compared to explaining the lessons to societies that did not want to know. The conservation of biodiversity became part of broader social, economic and political concerns. The science of conservation had to combine with campaigns for public awareness, participation and sustainable economics.

Evolving concepts for biodiversity conservation

Conservation science's conceptual framework shifted, even at the risk of it being submerged in wider topics. The United Nations Environment Programme summarises the need for this shift. (See Figure 5.1.)

The people must be able to:

1. Manage community involvement and participation

2. Identify water sources.

3. Identify land use areas (areas for wildlife, grazing, crops, people).

4. Protect natural resources.

Figure 5.1 *Zimbabwean CAMPFIRE publicity comic reflecting conservation's widening scope, combining protection of biodiversity, sustainable use and local ownership of resources*

Source: Zimbawbwe Trust.

1 To integrate different approaches to ensure the widest possible range of biodiversity is conserved.

2 To recognise that conservation is heavily influenced by social, cultural, economic and political factors.

3 To encourage co-operation and co-ordination of policy and institutions.

The conservation of habitats and species, in the sense of direct, science-based management is increasingly described as **protection**. Conservation now means much more besides. The World Conservation Strategy has three explicit components, protection, sustainable use of biodiversity and sharing the benefits of this use. Article 1 of the Rio Biodiversity Treaty echoes this with its remit as '. . . the conservation of biological diversity, the sustainable use of its components and the fair and equitable sharing of the benefits'. Protection of biodiversity is now allied with promoting sustainable exploitation.

Context

National policies and attitudes to conservation will affect the methods used. Success will depend on recognition of this and integration of conservation with national, social, economic and political goals. The UK and Zimbabwe are good examples of such differences.

The UK has a strong tradition of statutory and voluntary protection. The statutory mechanisms were organised by the 1949 National Parks and Access to the Countryside Act, revised as the Wildlife and Countryside Act (1981). The concentration on protection by use of protected areas is evident from the diversity of designations. (See Figure 5.2.)

- **National Nature Reserves**. These are nationally, often internationally, important sites, owned or managed by agreement with statutory conservation bodies. There are over 300 in UK.
- **Sites of Special Scientific Interest** (SSSIs or Areas of SSI in Northern Ireland). They were originally set up as counterpart to NNRs to provide regulation of valued sites outside the NNR system. They are now seen as main national mechanisms to safeguard sites, many outside formal reserves. In the UK there are over 6,000, with 8.1 per cent of land area.

Plate 20 *Elephants* (Loxodonta africanus) *in Zimbabwe at a water hole in front of a viewing platform in Hwange National Park. The impressive adults and playful babies are a fine sight but for local people they represent a direct threat to crops and life*

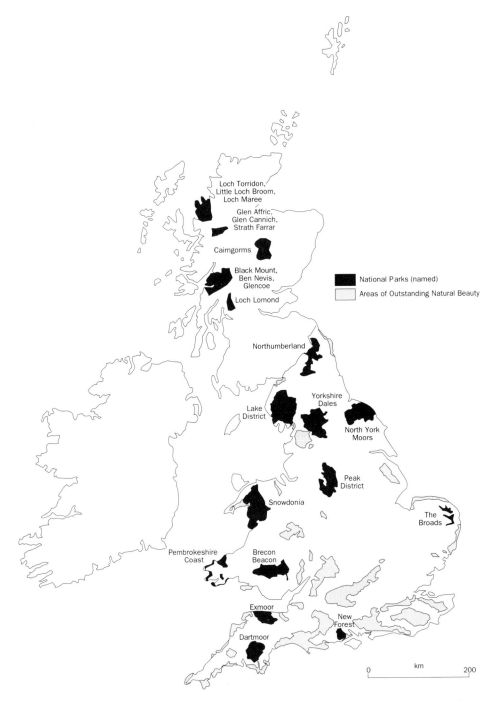

Figure 5.2 *Extent of two UK protected area systems. (1) National Parks (and their equivalent in Scotland). Multi-use protected landscapes encompassing the total area within their boundaries. Governing bodies have some restrictive powers and an increasing primary purpose of nature conservation. (2) Areas of Outstanding Natural Beauty, landscapes recognised for their natural and aesthetic quality. No single managing authority but within the area landowners may apply for aid to manage in an environmentally sensitive manner*

- **Special Areas of Conservation** (SACs). These sites have been selected, in response to EC Habitats Directive to maintain conservation value at a European level. In the UK 280 are proposed, many as areas embracing several individual sites assessed using SSSI criteria.
- **Special Protection Areas** (SPAs). These areas are designated for conservation of breeding and migratory birds in response to EC Birds Directive. There are over 70 in the UK. (**SACs and SPAs** are intended to form a pan-European network of sites, Natura 2000.)
- **Marine Nature Reserves**. (MNRs) These sites are much like National Nature Reserves but extend out to the limit of territorial waters. There are only two so far.
- **Local Nature Reserves** (LNRs). These are valued sites designated by local authorities. There are over 370 in the UK.

In addition to this protectionist reserve network biodiversity in the UK benefits from a host of other area designations that provide some regulation or incentive for good management: National Parks, Areas of Outstanding Natural Beauty, Environmentally Sensitive Areas.

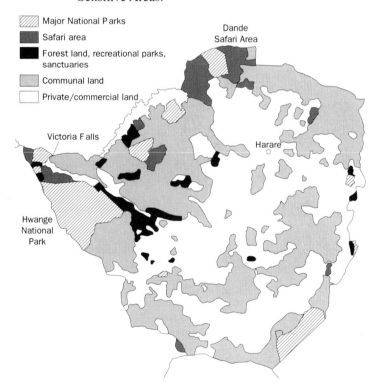

Figure 5.3 *Extent of Zimbabwean protected areas. National Parks, Safari areas and sanctuaries are all part of the Parks and Wildlife Estate controlled by the Department of National Parks and Wildlife Management. Communal Areas may take up the CAMPFIRE pro- gramme. The Dande Safari Area described in Chapter 4 is indicated as well as Hwange National Park (Plate 21) and Victoria Falls, Crocodile Farm (Plate 22)*

Zimbabwe's history has created a different structure. Responsibility for all wildlife is vested in the Department of National Parks and Wildlife Management, the infrastructure based on the 1975 Parks and Wildlife Act. The Act also set up a Parks and Wildlife Board to advise Ministers on conservation and utilisation of wildlife. Conservation is focused on the Wildlife Estate, over 13 per cent of Zimbabwe's area (See Plate 20.). The Estate is solely state-owned land or land held in trust by a statutory body. The Parks and Wildlife Act also recognises the work that can be done on private land and requires the DNPWLM to enhance wildlife management and production in all categories of land. Note the duality: conservation is linked with utilisation and production. The Wildlife Estate consists of five categories. (See Figure 5.3.)

- **National Parks**. These are intended to conserve the natural landscape, wildlife and ecology. While access for tourism and recreation is permitted the variety of activities is limited and sport hunting is excluded.
- **Safari Areas**. These areas are for the conservation of habitats and wildlife but with the express purpose of providing opportunities for public access for tourism and recreation, including sport hunting. Safari areas may be almost as valuable habitat as National Parks and are often adjacent. The difference lies in the policy of more extensive exploitation.
- **Sanctuaries**. These areas are set up to protect animals for the enjoyment and pleasure of the public. Sanctuaries are often smaller than National Parks or Safari Areas, supporting less diversity of species.
- **Recreational Parks**. These areas are set up for conservation of natural features for the enjoyment and recreation of the public.
- **Botanical Reserves** and **Botanical Gardens**. Botanical reserves and gardens are set up for the conservation of rare and indigenous plants and botanical communities.

Methods

Techniques for conservation fall into five main types:

- In-situ. Primarily the establishment and management of protected areas.
- Ex-situ. Maintenance of species in artificial environments, captivity, seed banks and the like.
- Restoration and rehabilitation. The reconstruction and repair of ecosystems.
- Land use management. Strategies to incorporate conservation in wider land use policies and practice.
- Policy and institutions. Appropriate financial, legal and infrastructure support.

Approaches

In order to mesh together the context and methods to cope with a complex world a number of approaches are utilised.

- **Strategic planning**. Conservation should not be an add-on gimmick but linked into the broader plans for resources, health, social and economic policy in a country. Article 6 of the Rio Treaty calls for National Biodiversity Strategies from all signatories identifying strategic needs within a general policy and Biodiversity Action Plans listing steps to be taken for selected species and habitats based on the National Strategy.

- **Bioregional management**. Goals set and actions taken must be pitched at a scale consistent with the ecological, social and economic forces at work. The piecemeal programmes of early conservation were too vulnerable to wider pressures. A first step is to identify bioregions, rather than working within often artificial national borders.
- **Adaptive management**. Our lack of knowledge coupled with nature's vagaries prevent certainty in our plans. Conservation must accept change and seek to work through it. This requires proper monitoring before and after conservation schemes are set up, feedback of results to refine management and an experimental approach so that results can be used to improve future work.

The shifting culture of conservation has had an impact. In 1994 the UK government published *Biodiversity: The UK Action Plan* which drew together existing policies, practice and strategy for the first time. This was a direct response to Rio's Article 1, an attempt to fuse ecological principles with economics and sustainability. Preliminary work driven by the Joint Nature Conservancy Council (the government's national umbrella) and the Department of Environment (a government ministry) identified four themes: conservation of resources (i.e. protection, establishment and management of areas); sustainable use; community involvement (from individuals through to government agencies); data collection and information. The UK Action Plan refined this into six aspects (conservation in-situ, ex-situ, sustainable use, partnership and education, UK support for biodiversity overseas, information and data). A steering group then took the UK Action Plan forward, producing *Biodiversity: The UK Steering Group Report. Volume I* in 1995 with detailed plans including:

- costed targets for protection of species and habitats;
- accessibility and co-ordination of biological databases;
- increase public awareness and involvement;
- mechanism for implementation and review of the Plan.

The need for a dynamic, updated and adaptive response is emphasised throughout. *Volume II* contains lists of key species, key habitats and broad habitat goals, costs and responsibilities. These include 116 key species and 14 habitat action plans are given plus reviews of the status of 37 main habitat types found in the UK. Species action plans identify responsibility for actions, rationale for targets, legislative protection, site protection, monitoring, review protocols, international co-operation, threats from air quality, climate and development of land. Habitat plans and statements include definitions, principal threats, rationale for targets and the broader context within which costed action plans can be drawn up. Boxes 19 and 20 summarise a Species Action Plan for Desmoulin's whorl snail and a Habitat Action Plan for Fens. The impact of biodiversity on conservation, with the horizons broadened beyond solely protection, had for the first time produced a national integrated species and habitat protection plan.

Box 19

Species Action Plan for Desmoulin's whorl snail,
(Vertigo moulinsiana)

The conservation status of some land snails, notably the *Partula* species of Hawaii, has become a *cause célèbre*. That Desmoulin's whorl snail, routinely described as 'about the size of a breadcrumb', should be shot to notoriety was not so obvious. However, this snail become a totem of the anti-roads protests staged against the building of the Newbury bypass in the UK during 1995–6. The protests, often involving direct mass action to obstruct the progress of clearing woodland, attracted sustained and often sympathetic coverage building on previous campaigns such as the M11 Twyford Down protests. As the protests reached their height it was suddenly realised that many of the wetlands across which the bypass would be built were home to this diminutive snail. At the same time the snail appeared as one of the 166 Species Action Plans. The Action Plan summarised: **Current status, Factors causing loss, Current Action, Action Plan objectives and Proposed action** (*Policy and legislation, Site safeguard and management, Species management and protection, Advisory, Future research and monitoring and Communications and publicity*).

Given the 'no action' proposal for publicity, Desmoulin's whorl snail attracted intense interest, television news coverage, newspaper articles and even a question on the BBC's prestigious 'Any Questions?' programme. An unfortunate representative of English Nature interviewed by the BBC resorted to stating that our lack of knowledge about the snail meant that its presence could not justify holding up the construction of the bypass, a reversal of the precautionary principle that probably sounded as foolish to him as it did the audience. The bypass is now being built.

The legislative framework

International treaties

Conservation laws developed as piecemeal, national legislation. International legislation conceded that sovereign rights were paramount and seldom recognised the importance of links in natural systems spilling over political boundaries. The Stockholm conference reiterated the importance of sovereign rights but also promoted responsibilities of states not to be a source of damage beyond their borders. The Rio Convention was the first international treaty to include legal obligations to act on this responsibility.

International treaties for conservation can work by imposing direct conditions but also through creating a supportive culture and incentives. Adoption of a treaty signals national support for conservation, perhaps a nudge to national legislatures to adopt specific laws. International treaties recognise global benefits and local costs of living with biodiversity, especially important to those developing countries harbouring the greatest diversity but least able to afford conservation or avoid destructive exploitation.

Box 20

Fens. A costed Habitat Action Plan

The Action Plan summarised: **Current status, Threats, Current action** (*Legal status, Management*), **Action Plan targets** (*identify priority sites in need of, and initiate, rehabilitation by 2005. All rich fen and other sites with rare species should be considered. Ensure appropriate water quality/quantity for all SSSI/ASSI fens by 2005*), **Action required** (*Policy and legislation, Site safeguard and management, Advisory, International, Future research and monitoring, Communication and publicity*) and **Costings** (the successful implementation of the Action Plan will have implications for the public and private sector. The Action Plan lists estimated costs to maintain 1,200 hectares through to 2010 (per annum): 1997 £40,000; 2000 £70,000; 2010 £70,000).

Building treaties

The first wildlife conservation treaty, The Convention for the Protection of Birds Useful to Agriculture, was signed in 1902. Since then over 150 environmental treaties have been signed, but lack of co-ordination, gaps and duplication impede their operation. Most treaties develop as a negotiated text adopted by countries but only in force once a country consents, usually by signature and ratification (or accession for later entrants). The treaty may only come into force when all participants, or an agreed threshold, consent. The success of a treaty typically requires review, update and administration to ensure compliance, through a secretariat and regular conference. Five of the global treaties prior to Rio have been the most influential for conservation. These are:

- The Convention on Wetlands of International Importance, RAMSAR. First signed in 1971, now adopted by 64 countries.
- World Heritage Convention. The Convention Concerning the Protection of the World Cultural and Natural Heritage, signed in 1972, currently adopted by 113 countries.
- The Convention on International Trade in Endangered Species of Wild Flora and Fauna, CITES. The first version was signed in 1973, now adopted by 122 countries.
- Convention on the Conservation of Migratory Species of Wild Animals was first signed in 1979, currently adopted by 40 countries, but only 11 full signatories.
- The Convention on the Law of the Sea, UNCLOS (50 in convention plus 107 signatories not yet ratified).

International treaties: mechanisms

International treaties use up to four approaches to conservation: **funding**, usually for national efforts on behalf of a global resource; designation of **protected areas**;

regulation of **biotechnology rights** especially in defence of local people; **regulated trade**, whether outright prohibition, restricted exploitation and mechanisms to ensure financial rewards to locals.

Funding

Funding schemes recognise the obligations of richer countries towards the developing world, so often required to conserve the bulk of biodiversity. Examples include the World Heritage Fund, linked to the World Heritage Convention (WHC). The WHC has proven a popular treaty due to the availability of finance and has an annual budget now in excess of $2 million. The Global Environmental Facility (GEF) grew from an IMF–World Bank Development Committee meeting in 1989 to finance protection of the global commons. Increasingly the GEF is a device to fund new international treaties. A three-year trial period started in 1991 with a $1.4 billion budget and has funded work to protect the ozone layer, control greenhouse gas emissions and conserve biodiversity. The GEF took on the financial administration of Rio Convention moneys, until a 1995 Conference of Parties (i.e. those countries that had signed the Convention) could agree on a mechanism.

Protected areas

The benefits of international treaties require signatories to designate protected reserves work via reciprocal rewards. A country may incur costs or forgo opportunities to establish reserves within its borders but benefits from similar sites set up in other states.

Biotechnology rights

Organisms, their parts and products were typically excluded from treaties regulating patents, ownership and exploitation rights but biotechnology and the potential of genetic diversity have propelled biotechnology rights to the forefront of international conservation law. In particular there is concern as to how to ensure that **source countries**, so often the developing world, benefit from exploitation of their biodiversity which is refined in the **recipient countries**, mostly the developed world.

Until the Rio Convention international treaties did allow some exclusive rights over plant varieties and other organisms. Plant varieties could be registered with the Protection of New Varieties of Plants (1961). Patents, designed for rights over new and useful inventions, were extended to organisms in 1980 in the USA as the result of a court case, Diamond versus Chakrabarty, over a genetically engineered bacterium. Patents now include genetically engineered mammals, famously the onco-mouse used in cancer research. The Rio Convention included no specific mechanism for patenting biodiversity. Articles 15 (Access to genetic resources), 17 (Exchange of information), 18 (Technical and scientific co-operation) and 19 (Handling of biotechnology and distribution of its benefits) all talk of exchange and fair and equitable sharing of benefits and co-operation, especially to help developing countries, a steer against restrictive patenting.

Tensions remain. Accusations of biopiracy have been fuelled by revelations such as those in 1996 that an American biotechnology company, Phytera, had signed contracts with several European botanical gardens under which they should provide tissue specimens from their collections. In return the gardens would receive a share of any profits from products developed. Unfortunately some of the contracts omitted commitments to share some of the profits with the countries of origin of the plants.

International trade in wildlife

The total value of wildlife trade (whole organisms, parts and products) may equal that of the forestry or fisheries industries. Trade is regulated by the CITES, originally a prohibitive treaty, forbidding all trade in listed taxa, Appendix I, and trade by permit only for others, Appendix II. CITES also requires signatures to provide annual records of Appendix II trade, to monitor trends and check against estimated stocks. CITES emphasis has shifted. The 1981 New Delhi Conference of Parties permitted change of some Appendix I species to Appendix II for sustainable use, primarily ranching. The first example was the Nile Crocodile in Zimbabwe, 1983. A Management Quota System, relying on countries to devise quotas and consumer countries to enforce compliance accepting only those imports with valid documents, was permitted. These opportunities brought problems. National quotas could be poorly set, either by ignorance or design. Even if a country's quota was reached wildlife could still be harvested, smuggled over the border and laundered into stocks of adjacent states.

Trade can create positive results. There is a direct financial reward in return for conserving biodiversity, ideally filtering down to local people. Proper trade has encouraged more controlled captive ranching, e.g. butterfly farming in Papua New Guinea, green iguanas in South America, crocodiles in Zimbabwe. (See Plate 21.)

There are currently some 675 Appendix I species, 3,700 Appendix II animals, 21,000 plants (though whole families can be listed, so the inclusion of the orchid family adds up to 35,000 species).

The Convention on Biological Diversity

Rapidly evolving biotechnology and trade treaties reflect the ecological, economic and social origins of biodiversity. The Rio Convention displays

Plate 21 *The author wrestling with a would-be Zimbabwean handbag at a crocodile farm at Victoria Falls. The farm hatches eggs laid in captivity and 5 per cent of two-year-old crocodiles are released into the wild. The rest are turned into clothing, ornaments or meals*

this mix of conservation by obligation, equity rights and positive incentives. Access to genetic resources by foreign collectors requires prior consent by source countries, fees and royalties and mechanisms to share the outcomes of research. The transfer of technologies for conservation to the developing world is not only encouraged but requires preferential terms. This equity and access by source countries and people overrides patent laws. Essentially conservation in the developing world is explicitly tied to obligations on the developed world, for expertise and money, despite the best efforts of some signatories to write opt-outs into the Convention treaty (Box 21).

Box 21

Rio Convention caveats

Articles 20 and 21 of the Rio treaty deal with financial resources and financial mechanisms respectively. Paragraph 1 of Article 20 states 'each Contracting Party undertakes to provide, in accordance with its capabilities, financial support and incentives in respect of . . . the objectives of this Convention', followed by paragraph 1 of Article 21 'There shall be a mechanism for the provision of financial resources to developing country parties'. The UK's official publication of the treaty includes this revealing caveat at the end: 'The government of the United Kingdom . . . declare their understanding that the decisions to be taken by the Conference of Parties under para. 1 of Article 21 concern "the amount of resources needed" by the financial mechanism, and that nothing in Article 21 authorises the Conference of parties to take decisions concerning the amount, nature, frequency or size of the contributions of the Parties'. France and Italy added similar nervous notes.

National legislation

National legislation differs in detail between countries but common themes recur. Laws can be categorised into two approaches: first, regulatory, involving prohibition or restriction, backed by punishment versus non-regulatory relying on positive incentives; second, the focus for laws on genetic resources, species, habitats or destructive activities.

Regulation

Genetic resources

Legislation to conserve genetic resources is a recent development, with little precedent to build upon and an indicator of changed attitudes to conservation. Protection is typically combined with regulation of exploitation and sharing the benefits. Regulation of genetic biodiversity is founded in the Rio Convention's statement of the sovereign rights of nations to control access and use of genetic resources, vividly revealing the tensions between source and recipient countries. Source countries face three tasks: regulation of collection, including sharing information and participation of local people; regulations for sharing technologies and benefits; establishment of regulatory authorities to oversee compliance. In addition source countries will only succeed if recipient countries develop complementary legislation to allow enforcement.

Species protection

Traditional custom, taboo and law protected some game animals in many countries, with restrictions by season, catch, method or entitlement. The majority of species went unprotected, often held to be the property of the landowner. The first species specific law was drawn up in the Swiss Canton of Zug in 1911 to regulate collection of the Edelweiss, *Leontodon alpinum*. Article 8(k) of the Rio Convention calls for countries to develop legislation for the protection of threatened species.

Species based legislation commonly lists chosen species, banning or limiting interference. The taking and possession of wildlife are often linked to bans on trade. Since the act of collection often goes unseen the prohibition of possession, whether of whole specimens or parts of products, is the best control and trade restrictions limit financial rewards.

Legislation can also require conservation and recovery schemes, again for a species or habitat, financial incentives and, rarely, the precautionary principle that any collection or alteration of habitat must be proved to cause no harm. The Sri Lankan Fauna and Flora Protection Ordnance 1993, is an example of the latter.

Problems for listing species arise from lack of criteria, evidence or expertise. Many national lists are short, dominated by the inevitable spectacular vertebrates. Legislation is easier to draw up for state-owned land. Elsewhere restrictions on the rights of private landowners may be regarded as an infringement of liberty and cause resentment.

Protected areas

The bland jargon 'protected areas' covers a host of legislation from nature reserves, characterised by prohibition or restriction of activities within designated sites, habitats or regions and works in many ways.

Specific nature reserves owned by government or voluntary conservation bodies are the most obvious. State control of land can be problematic, requiring compulsory purchase or first refusal to buy and, if these are acceptable, may be limited by money. Management can be impeded by the need for buffer zones, external impacts that cannot be escaped, e.g. atmospheric pollution, and the human problem of integrating local interests with what might be seen as the heavy hand of government or outsiders.

Conservation can be achieved without the rigour of a protectionist reserve. Many countries designate **wilderness areas** that forbid and regulate certain activities by landowners and visitors, the land use controls perhaps zoned to allow some activities in parts but not throughout. Other approaches include site-specific regulation of privately owned land, e.g. UK's SSSIs or are habitat specific, granting general protection to a habitat type throughout a country whether an ecosystem, e.g. wetland, or more particular feature, e.g. pond. Planning controls covering all land can be useful, especially inclusion of less valuable landscape, e.g. urban. These controls are either comprehensive, anticipating developments, perhaps through zoning, or reactive, dealing with applications on a site-by-site basis.

Regulation of damaging activities

Many human activities degrade biodiversity, their impact spreading far beyond the source. However well established a protected area is these outside activities can be a serious threat and national legislation often seeks to limit such damage. There are two main problems, activities within a protected area that unintentionally cause damage and those causing damage beyond their source site. The former includes the impacts of walking, climbing and vehicle scrambling. In such cases the very quality of the natural habitat is the lure. The second category includes pollutants, pesticides and the release of exotic species. Damage from either type is difficult to regulate. Legislation often results from damage already done. Damage on private land is especially tricky to control. Incentives for good management may be more effective.

Non-regulation. Incentives and agreements

Regulation can be ineffective, stymied by lack of means or the will to enforce. Positive incentives to conserve are increasingly used. There are three main forms. Financial schemes include incentives to conserve such as tax reductions or exemption for conservation management schemes and abolition of incentives to destroy. Restrictive agreements require a land owner to manage a site in a specified way. Such agreements include covenants which work in perpetuity so that subsequent owners must maintain the work, and easements which are an obligation often linked to a very precise task or action. These agreements sound coercive but can allow conservationists to sell land that can be used in some ways so long as the specific goals are met. Management agreements use contracts to start or maintain positive management in return for payment. The payments can be linked to a particular practice e.g. low intensity farming or creation and maintenance of specified habitats.

International aid

The Rio Convention explicitly stated the need for financial aid for conservation from the developed to the developing world. Richer countries were asked to back their presidents' speeches with money. Financial aid is a vital complement to laws and treaties.

Development aid

Biodiversity can benefit from unilateral development aid, i.e. from one country directly to another, but specific allocations to conservation are difficult to pinpoint in general aid budgets. Of the sixteen countries in the OECD Development Assistance

Committee, in 1988 only six could identify funds allocated specifically for biodiversity, though others were moving towards this. In addition to direct funds the screening of other aid projects for likely negative impacts is also possible. Of the sixteen OECD countries, eight did so with more to follow. Funds also flow via multilateral aid, through sources such as the World Bank, co-ordinated by the Committee of International Development Institutions (CIDIE). Precise allocations to biodiversity are difficult to isolate and screening of all projects for impacts was slow to develop, with the World Bank introducing **environmental impact assessments** (EIAs) in 1989.

Clashes between different parts of the World Bank still occur. The CIDIE was also the first administrator for the Global Environmental Facility (GEF) which was set up in 1989 specifically to fund biodiversity and environmental conservation. In its first three-year plan (1992–95) US$4,000 million was allocated to biodiversity projects from a total budget of US$1.4 billion. Between 1992 and 1997 US$250 million was available annually. The GEF has suffered problems. Initial projects were classic protectionist conservation, with little feel for wider contexts, local participation and benefit. The lack of participation may be due to confidentiality of bank procedures. The fund also concentrated on large short-term projects, e.g. establishment of big reserves, rather than long-term, local scale needs in developing countries.

Recently National Environment Funds have been promoted as a way forward. Local, national and international expertise can combine. Such bodies could facilitate the flow of international funds, combining local expertise and control over what might be resented as foreign interference. There would also be accountability to donor countries and the chance to regulate the boom-time bonanzas and cash-flow starvation that can afflict conservation efforts.

Overall estimates for recent (1994) protected area budgets are US$4.1 billion, a lot of money but far short of the US$1 trillion for defence, US$245 million on agricultural subsidy and a long way short of the US$17 million that the Global Biodiversity Strategy estimates as the annual need for effective management of global biodiversity.

Tropical forest aid schemes

The 1980s focus on tropical forestry lead to the foundation of two schemes in 1983: the Tropical Forestry Action Plan (TFAP), combining the UN and conservation NGOs, and the International Tropical Timber Agreement (ITTA), a cartel of producer and consumer countries. Although seen as rivals, especially given the potentially different agendas of their founders, their actions have been similar, including protection and transfer of skills and benefits. The TFAP co-ordinates aid for the conservation and sustainable use of forests, based on National Forest Management Strategies for each country. Funding for advice, followed by review of resulting plans, creates agreed strategies which help donors to channel aid into recognised projects. Forestry conservation funding rose from US$603 million in 1984 to US$1,080 million in 1986 at a time when overall aid fell in value by 50 per cent. Biodiversity conservation is just one focus of the TFAP, taking 8–9 per cent of the budget,

US$83.5 million in 1988, with a target of US$1,550 million over the five years 1985–1990. The ITTA had the hallmarks of a commodity agreement but from the start recognised conservation as part of utilisation to maintain ecosystems and biosphere. The total ITTA budget in 1991 was $20.4 million, of which the committee funding forest management plans, the mechanism used to build in conservation, had $12.3 million. With the grand aim of sustainable management of tropical forest by the year 2000 ITTO has published 'Guidelines on the Conservation of Biological Diversity in Tropical Production Forests'.

Debt for nature swaps

Debt for nature swaps convert part of a country's financial debt into a domestic obligation to conservation. Precise details vary but the main steps are broadly similar. Debt for nature swaps seldom involve simply wiping out a debt in return for a promise to conserve. Inside a conservation organisation must raise money to buy the debt from the bank to which it is owed. Since the debts of so many developing countries are never likely to be paid off, the conservationists can buy the debt at a fraction of its real value, often 15–30 per cent of the amount owed. The debt is bought in a hard currency. The conservationists can then negotiate with the debtor country to fix a rate at which the hard currency can be converted into local currency, called the **redemption price**. The debtor country then issues a **bond in local currency** covering the redemption price. The bond can be used to finance projects. All parties stand to gain. The banks get some repayment, the country is extricated from some debt, conservation projects can be established.

Too good to be true? Swaps can be seen as an imposition with foreigners controlling a country's debts but since the debts are originally owed to foreign banks and the swaps result in local control of finance this is less of a danger. Swaps can risk imposing their ideals and projects but sensitive handling can create a positive link helping local communities. More awkward are problems with governments that are unwilling to see power and money available disseminated at a local level. Debt swaps may set off unexpected market forces, creating local inflation. At the international level the prospect of debts being bought out has raised their redemption value. As more and more are purchased the value of those left rises. Redemption values of several South American countries' debts may have rallied for no other reason than the intervention of conservationists, making debt swaps more expensive. So far debt for nature swaps have redeemed less than 10 per cent of the debt burden, risking a false impression of success while unsolved debt crises continue to drive the ultimate human pressures that are the major threat to biodiversity.

Protected areas

Protected areas is the catch-all term for what many think of as nature reserves. The shifting emphasis of conservation away from purely protection of wild nature has stretched the definition of reserves. The Rio Convention defines a protected area as a geographically defined area which is designated or regulated and managed to achieve specific conservation goals; the Global Biodiversity Strategy as legally established land or water area under public or private ownership that is regulated and managed to achieve specific conservation goals. These broad definitions recognise the variety of purposes conserving genetic, species or ecosystem diversity and, increasingly, the importance for human welfare and culture, for sustainable use and integration of multiple uses such as recreation and harvesting. What unites protected areas is regulation of human impacts, within a defined area, established by legislation or ownership and with specific goals. As UNEP wryly puts in 'packing devices to achieve defined ends'.

This approach to conservation is very ancient, in sacred groves, hunting parks and wilderness. Themes redolent today are apparent in these ancient schemes, the desire for wilderness, perpetuation of wildlife, regulation of exploitation, protection of forests. (Table 5.1 lists examples of historic protected areas.) The modern concept of reserves was a nineteenth-century evolution of these strong foundations.

Table 5.1 *Historic examples of protected areas and allied laws. Note the similarity to modern themes via establishment of areas and regulation of activities*

Protected area	Purpose
Middle Kingdom Egypt	Hunting licences to regulate wildfowling.
Sumero-Babylonia laws 1900 BC	Fines for cutting down trees.
Wessex Law Code (Britain) 700 AD	Penalties for cutting down trees.
Norman Britain eleventh century	'Forests', a specific area designated for hunting exclusive to the king.
Henry VI (1423–27)	Laws to regulate netting of river fish and control damage to coastal marshes and sea banks.
Victoria (1860–92)	Eight Acts of Parliament dealing with game licences, protection for wild birds and poaching.

The purpose and role of protected areas

The use of protected areas for conservation recognises that habitat loss, fragmentation and degradation are major causes of extinctions. The long-term survival of many species and ecosystems outside of protected areas in landscapes exploited by people is unlikely. We do not have the expertise or resources to maintain most species ex-situ, in zoos, botanical gardens or gene banks. We do not even know how many species there are, so protecting areas hopefully includes the unknown as well as the familiar. Reserves also nurture aesthetic and cultural attraction, a wild idyll of landscape, often the spur to conservation centuries before the concept of biodiversity was coined. Protected areas are a fundamental device for conservation in all countries, though the numbers, areas and goals vary with landscape and culture. Historically protected areas have arisen by a serendipity of social mores, money and available land. The recognition of the true extent of biodiversity, the hidden iceberg of species numbers, genetic variety, ecosystem services and future insurance submerged beneath the historical focus of a few game species and myths of wilderness, has resulted in attempts to distil the goals of reserves and strategies for selection, design and management.

Protected areas used to be seen as isolated pockets in the wider, exploited landscape, pockets that could somehow be managed to retain their original character with little regard to the world beyond. But wire fences cannot exclude the economic pressures, global change, pollution or local people. Effective use of protected areas now recognises the importance of ecological processes and change. Reserves should not be set up which are too small, fragmented or isolated so that their ecology is not sustainable. Reserves must account for evolutionary processes and genetic variety. Reserve management also commonly includes use by people.

Protected areas for biodiversity commonly cite four main goals for conservation:

- functioning ecosystems, recognising that processes which are being guarded may themselves drive change;
- biodiversity, e.g. species richness, endemics, distinctive wildlife, often requiring management to prevent change to maintain the desired state;
- species specific, typically charismatic, flagship species;
- exploitation and sustainable use, recognising human benefit now or future potential.

Selection, design and management

Selection, design and management are linked. Selection results from the strategy, the ideals, the techniques used. Design is a local response to available choices. Management is the plan for the future.

Selection

The threats to biodiversity have prompted a more rigorous approach to selecting reserves than the accidents of history. Recent approaches combine **area** and **representativeness**. Ideally national boundaries are less important than selection within a **bioregion**, though this may be politically impossible. A larger area will harbour more species, more communities and have more robust ecological processes. Any selection strategy should aim to include the full range of taxa and habitats in a country. Selecting representative areas requires some **classification** of ecosystems. Selection by area and representativeness should then be reinforced by identification of **gaps**. This is especially important for a country with historic reserves that are the result of an ad hoc focus on rare and special species and habitats and often ignore examples of the common. In addition **hot spots**, defined by species richness, endemism or distinctiveness, can be added. Selection for ecologically viable area, representativeness, filling gaps and hot spots presumes that expertise and data exist and reflect the intrinsic value of biodiversity.

Other criteria can be useful. Reserves selected for **flagship species**, typically large vertebrates, are common. An area sufficient in resources and large enough for a viable population of animals such as these protects many other species, even if little known. The symbolic value and perhaps tourist income add political and financial benefits. **Indicator taxa**, usually species readily found and identified but known to reflect wider biodiversity can be useful where data are limited. Wilderness, defined almost by our lack of knowledge, certainly by lack of human impact, has been used on a global scale. Three global wildernesses have been proposed: one an arc through the Amazon; second, the Zaire basin; third the New Guinea rainforests. Areas of this size provide global services and an evolutionary pool. **Utility** may be used to safeguard sustainable exploitation or future potential, e.g. conserving wild ancestors of crop species as an insurance. Even if very little is known of a region the ecological principles outlined in Chapter 3, particularly the patterns associated with broad primary factors of area and gradient, may be used to designate sites.

Design

Protected area design has been a fraught topic. In developed countries any chance for design is limited by the scramble to protect remnants, whatever their size and shape. Design from scratch is only possible when carving out new areas in wilderness. Design has become snarled up in a debate between theory, largely based on **island biogeography theory** since reserves are effectively islands in a sea of degradation, and practice. Some sound basic principles do exist.

Size and number

A larger area will often be a more successful reserve, harbouring more biodiversity. Larger populations are more resistant to extinction and larger species often require wider ranges. The amount of edge and fragmentation will be less relative to the core and the whole will be less vulnerable to a single natural disturbance (at least of a kind characteristic to the habitat, e.g. fire or flood). Essentially the area must be large enough to be a **minimum dynamic area** for the ecology. The size must be sufficient to provide the resources required to support the **minimum viable population** of a species, whether rarity, flagship or keystone. So big is best is a simplification. All depends on the size of species and nature of habitat in question. The size needs to be large enough to avoid accidents (demographic e.g. not finding a mate, being eaten or environmental, e.g. fire, flood), genetic or social meltdown and to permit migration and ecosystem function.

An equal area divided between several small reserves may be more appropriate than one big site, the **Single Large Or Several Small** (SLOSS) debate. Several reserves may be less at risk than one larger site hit by disease, predation or severe disturbance because the risk is divided. This is especially so if the ecology of a species is out of sync between different reserves. For example, if one stage of a species' life cycle is vulnerable to disease or disturbance, populations in all reserves will be not be at this same stage. Surviving reserves act as recolonisation sources. A scatter of reserves may hold more species in total because they can cover more variety of habitats. Genetic diversity can be maintained. Size and number vary with the goals of the protected area.

Shape, fragmentation, buffers and zoning

The shape of a reserve will determine the length of **edge** relative to area. Extensive edges have been seen as bad, allowing protected species to leave the safety of reserves, entry for exotics, human pressures and pollution. **Fragmentation** increases edge length and creates other problems. Populations may be cut up into smaller, less viable pockets. Barriers to dispersal can either be physical, e.g. fences, or behavioural barriers, e.g. open roads that animals will not cross. Circular, intact areas have been promoted as the ideal minimising edge to area. Sprawling shapes, especially if long, can cover more habitats increasing diversity. Edge habitats can also be valuable, either due to intrinsic combinations of species or as foraging habitat. The greater problem is that the edge of any reserve is not inviable to human impacts.

Design incorporating the insights of the SLOSS debate and shape should see a protected area within the wider landscape context. Reserves can be established in **sets**, recognising distances over which species can move. **Buffer zones** can be created. Many protected areas cannot be isolated from the surrounding land, bringing problems such as agricultural fertiliser run-off in Britain or local people grazing livestock and cutting wood in Africa. New reserves may have to accept historic use and future pressures. Increasingly reserves are created that accept multiple uses and use zoning in place of abrupt edges. Zoning allows different activities within defined parts of the reserve. This also helps build an infrastructure to suit the reserve. UNESCO Man and Biosphere reserves use this approach. The core area is totally protected with a surrounding buffer permitting traditional use and an outer transition zone allowing sustainable development and experimental management. This approach recognises problems rather than taking an oppressive conflict stance.

Corridors and connections

Linked to size and shape are ideas for routes, whether open corridors, linear habitats, stepping stones or flyways to help movement between reserves. Dispersal includes daily routines, annual migration, dispersal stages and shifting overall range. Corridors have been promoted to permit recolonisation of isolated reserves and extend the effective ranges of smaller areas containing large, mobile species. Corridors also bring threats by allowing spread of disease and exotics and providing a focus for poaching and increasing edge. As so often the answers are species specific.

Position in landscape

Reserves cannot be designed in isolation from the surrounding land or seascape. They must be seen as part of the wider ecology, including human land uses. A regional approach can incorporate sets of reserves, buffers, zones and multiple use from the start. Part of the success of a protected area will be the effectiveness of management beyond its boundaries, requiring a national and regional coherence as well as local actions such as education.

Management

The best selection and design protocols will be a waste of time without effective **management goals**. Protected areas need clear goals and effective management to realise them. This is a matter of political will, expertise and money. Areas with unclear or conflicting goals will fail. Management for one species or habitat may threaten others. Unclear responsibilities and conflict can disrupt progress. Willingness to adapt in response to research and experimental management, perhaps to alter the size and shape of a reserve can be important. Involvement of local communities in decision-making, financial benefits and management of costs, e.g. damage to crops or dangerous wildlife, can all ameliorate pressures.

Protected area problems

Not all protected areas are successful. Management failures have occurred on many sites. Very successful management can also create threats by imparting a false sense of security and providing symbols of success that none dare alter. This attracts increased pressure from tourists, poachers, local people and industry, all keen to exploit pockets of wildlife. Local conflicts are commonplace where protected areas try to operate an exclusion policy. Recent shifts in conservation policy, e.g. emphasising benefits to and involvement of local people, have highlighted the lack of sustainable use of biodiversity resources inside many reserves. Utilisation may be difficult to build into existing management plans. There are problems of bias of designated sites, globally and nationally. Some ecosystems are extensively protected, others underrepresented and establishment of marine reserves lags behind terrestrial habitats. Illegality can create effective protected areas. The very rare Lears macaw, thought restricted to 177 individuals at one site in South America, has recently been discovered at a second, high quality habitat site. Bird trappers and traders have left the birds alone for fear of drug cartels growing marijuana around the site.

The extent of global protected areas

The IUCN Commission on National Parks and Protected Areas acts as scientific and technical advisor for selection, establishment and management of protected areas, with the CNPPA and WCMC jointly monitoring progress. Recognition to join the CNPPA list requires the proposed area to meet strict criteria: First, size – only sites larger than 10 sq. km (or whole islands of more than 1 km^2) are included. Second, the area must fit one of the recognised eleven IUCN management strategy types, of which I-V are typically regarded as true protected areas, the new category XI also fitting this designation (Box 22 outlines the IUCN categories). Third, the site must be controlled by the highest authority possible within a country, usually central government though federal state level may be accepted.

Box 22

IUCN Reserves categories

I Scientific reserve/strict nature reserve To protect and maintain natural processes to maintain natural ecological and evolutionary systems.

II National Parks To protect outstanding natural scenic and natural areas, including recreational and educational but not extractive use.

III Natural Monument/Landmarks To protect nationally significant features of special interest or unique character.

IV Nature Conservation Reserve To protect nationally significant species, communities or features requiring human intervention. Some exploitation permissible.

V Protected landscapes/seascapes To maintain significant land and seascapes characteristic of harmonious interaction of humans and nature.

The remaining categories are often separated from the true protected areas of I–V.

VI Resource Reserve Relatively undeveloped areas under pressure for development. Human impact not poorly understood.

VII Anthropological Reserve Natural areas where modern impact has not significantly interfered with or been absorbed by inhabitants.

VIII Multiple use areas Large areas containing significant natural features and systems but also suitable for exploitation.

The IUCN also categorise sites designated by two major treaties.

IX World Heritage Sites

X Biosphere Reserves The changing focus of conservation, away from protectionist policies has seen the introduction of another category.

XI Managed Resource Protected Area Protection of biodiversity while providing traditional sustainable utilisation by local people within a largely unmodified area.

By these criteria some 10,000 sites are now recognised, from 169 countries and covering 5–6 per cent of the Earth's surface. This is less than a third of the 37,000 protected sites listed by the WCMC, many of which do not meet the criteria. (See Plate 22.) Note the requirement for control by the highest authority also excludes many reserves owned by voluntary bodies. (See Figure 5.4.) Numbers of protected area reserves and allied designations globally and in Europe and Africa are given in Table 5.2.

Besides the classic problems of underfunding, poor management and constant threats of exploitation the increasing co-ordination of global reserve resources has shown big differences of biogeographical coverage. Of sub-Saharan Africa 5–7 per cent is designated within category I–V reserves compared to only 0.9 per cent of Europe. Political boundaries are less useful than **bioregions** or **biomes**. The extent of protected areas differs across the classical biogeographical realms, from 242 sites covering 0.59 per cent of Antarctica to 1920 sites over 12.4 per cent of the Nearctic. At the biome level protection ranges from 194 sites taking in only 0.8 per cent of temperate grasslands up to 935 sites representing 9.3 per cent of subtropical and temperate rainforests.

Plate 22 *A campsite in a Zimbabwean National Park. Permanent structures provide some refuge for tourists in an emergency but tents can also be pitched. The site is managed by the Department of National Parks and Wildlife as part of its conservation work*

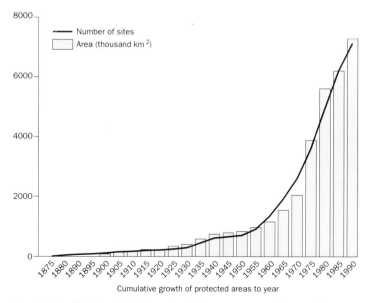

Figure 5.4 *Global extent of protected areas measured by number and extent*

Source: Redrawn from Groombridge 1992.

Ex-situ conservation

Protected areas are the main device for in-situ conservation, relying on ecosystems to sustain themselves, with some help. An alternative for endangered species is to establish a captive population away from its natural habitat. This ex-situ approach includes **captive breeding**, **release** schemes and **biodiversity banks** such as seed collections. Several species have famously been saved from extinction by ex-situ schemes. Some species still only survive in captivity. The earliest known example of an ex-situ success is Père David's deer, extinct in the Chinese wilds some 3,000 years ago but surviving in a park. Other celebrity examples include the NeNe goose of Hawaii, nurtured at the Wildfowl and Wetlands Trust, the black footed ferret in the USA, the entire remnant population taken into captivity between 1987 and 1991 to escape canine distemper, and the *Partula* snails of Hawaii, a rare example of popularised invertebrate conservation due to the survivors living in sandwich boxes at London Zoo.

Table 5.2 *Numbers of IUCN category I–V, IX, X and RAMSAR protected areas globally, in Europe, Africa, UK and Zimbabwe*

Region	I	II	III	IV	V	IX (World Heritage Site)	X Biosphere (Reserve)	RAMSAR
						IUCN management category		
World	712	1,483	323	4,021	1,952	95	300	538
Europe	115	137	43	712	1,143	15	91	328
Africa	40	222	2	405	29	29	43	43
UK	3	1	0	45	110	3	13	45
Zimbabwe	1	12	1	5	3	2	0	0

Source: Groombridge 1992.

Ex-situ and in-situ conservation should not be thought of as alternatives or rivals. Ex-situ schemes should be linked to in-situ projects to augment remaining populations or establish new ones. Ex-situ schemes can link directly with protected areas, providing stock to release, research data and, in the case of zoos, funds from ticket sales. Ex-situ populations are a refuge while in-situ threats can be dealt with. Captive stock create additional benefits as popular icons, inspiring interest and finance. They act as flagships and are important for education, spotlighting endemism, hot spots or particular threats. Successful release schemes are effective symbols, building more support for conservation. Wild populations will be important to captive stock to maintain genetic diversity and reinforce captive bred stocks. Ex-situ programmes raise problems equal to their benefits.

Whatever the precise number of species on Earth, 10, 20 or 30 million, the resources and technology for ex-situ conservation of all of them do not exist. There is a bias to mammals, birds and some plants. Of the 20,000 terrestrial vertebrate species estimates suggest that only 1,000 species could be sustained ex-situ. Some taxa are probably impossible to maintain in captivity, e.g. the large whales. For those species that can be supported the effort may be expensive and risky, the money better spent in the wild. Costs of maintaining African Elephants, *Loxodonta africana*, and Black Rhinos, *Diceros bicornis*, in zoos may be fifty times that of protecting the same number in the wild. Nurturing tiny populations in captivity can create a false sense of security. Evolutionary and ecological variation found in separate wild populations of a species may be lost, so genetic and ecological biodiversity decline even if species diversity is maintained. Adding to captive stocks can deplete remnant wild numbers. Captive populations are typically small and scattered, risking genetic and social problems. Successful breeding brings its own problems. Captive born young may be naive, partly domesticated individuals that cannot be released: The redemption of a species in captivity may make exploitation of its wild habitat politically easier once it is apparently secure ex-situ. Not all captive stocks have been successful. Some species have met their doom in captivity, e.g. the Passenger Pigeon, though to be fair there was only one bird left. More recent schemes have foundered. The Sumatran Rhino, *Dicerorhuns sumatrensis*, was considered so endangered in the wild that a captive stock was considered essential; 38 animals were caught depleting wild populations but 15 died and none have bred. The ethics of captivity, particularly the role of zoos, have provoked criticism with suggestions that captive stocks are more to the benefit of zoos than the animals. Extinction with dignity in the wild has been touted as preferable to miserable captivity. Even if all the objections were overcome the ultimate problem remains. If no viable habitat is left in the wild, no release is possible.

The variety of past schemes and lessons from success and failure have driven attempts to codify principles and practice for ex-situ conservation. Ex-situ schemes are now seen as support for populations in their natural habitat. Ex-situ projects should be used as refuges for populations in immediate danger of extinction, as part of programmes to ensure the long-term survival and for education and research. Projects should promote the longevity, health and quality of life for captives, ensure reproduction maintains stocks to avoid taking more from the wild and maintains natural patterns of genetic diversity, sex ratios and age structures. Captive breeding can then support releases, re-establish ecologically or culturally important species and perhaps bring economic benefits. These goals need an understanding of the biology of small populations, the experience, training and technologies to carry them out and management to co-ordinate work, not just with the captive stock but responsive to changes in wild populations. The result is international co-ordination of ex-situ schemes, controlling both the management of captive stocks and attentive to wider priorities for species conservation.

The IUCN/World Conservation Union have a Species Survival Commission to which the Captive Breeding Specialist Group (CBSG) reports on individual species. The International Species Information System (ISIS) uses tools such as the Animal Records Keeping System (ARKS) and studbooks which record the ancestry and whereabouts of individual animals. ISIS handles data on 4,200 animals, from 395 zoos in 39 countries. There are over 300 breeding programmes, featuring over 200 endangered species. Co-ordination extends beyond cataloguing existing schemes. The CBSG and Taxon Advisory Groups use Conservation Assessment and Management Plans (CAMPs), which include Population and Habitat Viability Assessments (PHVAs) to identify priorities and future needs or opportunities. CAMPs are strategic plans for management of threatened taxa. Workshops collate existing data and assess threats and extinction probabilities and population estimates for both in-situ and ex-situ stocks. The species is then assigned to a category (critical, endangered, vulnerable or safe) and recommendations are made for action. These can include the need for PHVAs, management plan workshops, conservation ex-situ, gaps in research and technology or captive breeding. The outcome of a CAMPs review is a Global Captive Action Plan (GCAP), an international or regional plan for captive breeding but in support of in-situ conservation. GCAPs assess which species should remain in captivity, be added or need not be, and review available resources and technology plus research and financial benefits to wild populations.

Captive populations

Captive breeding or raising is central to ex-situ conservation in order to buy time for in-situ projects. Captivity can be a last resort in the face of imminent extinction but ideally captive populations should be established before this, allowing time for research into species' needs, causes of decline and to ensure that taking from the wild does not add to losses. The possibility of success, eventual release and cost compared to in-situ protection are important. Use of studbooks to track individual family trees helps to minimise inbreeding and positively to increase genetic diversity via planned matings. Where transportation of animals is physically difficult artificial insemination, egg and embryo implants might be used.

Captive raising includes **cross-fostering**, generally incubation of eggs by domestic birds, though care must be taken so that chicks do not imprint on their surrogate parents on hatching, growing up believing they are the same species. **Head-starting** is protection of wild eggs and help for hatchlings e.g. of turtles and crocodiles, to overcome heavy mortality at hatching, either from predation or sometimes disorientation and obstacles created by human development of beaches. Although superficially effective, such schemes cause problems. Young born in captivity may lack vital skills so release is impossible or, conversely, retain the very skills and behaviours that doomed their in-situ forebears. Head-starting may be pointless since natural populations of the species involved typically withstood severe mortality of hatchlings, the threats causing population declines occurring at other stages in the life cycle.

Plate 23 Partula *snail reintroduction. Enclosure built on the Pacific Island of Moorea to product reintroduced* Partula *stocks. The electrified walls are intended to exclude predatory* Euglandinia *snails. Reintroduced* Partula *survived initially and bred but* Euglandinia *got in over branches falling across the walls wiping out the* Partula

Photograph Dave Clarke/London Zoo.

Releases

Captive breeding hopefully leads to releases, often called **translocations**, which take several forms. **Augmentation** (or **restocking**) is the addition of captive bred or wild caught individuals into an existing population. **Reintroductions** (or **re-establishment**, **restoration** and **repatriation**) establish new populations within a species' historic range, in areas from which it has been lost. (See Plate 23.) This could be back into original habitat from which the captive stock was taken or elsewhere in the range if suitable sites exist. **Introductions** are releases outside the original range. These can be a threat to the host ecosystem. Wanton introductions have been a major cause of extinctions as exotic species flourish. Even those undertaken for conservation have been pilloried as unnatural gardening. International translocations of vulnerable species, e.g. attempts to establish black rhinos in Australia, beyond the reach of poachers are acceptable examples.

Releases can bring many benefits. The likelihood of extinction in the wild is reduced, ecosystem functions can be restored, especially if the species is a keystone, and the risk of catastrophe and inbreeding to captive stocks are ameliorated. Successful schemes can act as potent flagships for conservation. However, releases are almost certainly doomed if the original causes of decline and extinction remain. Given the long expensive and difficult nature of releases some principles have been developed from analysis of successful schemes.

The release area
The original cause of decline and any new threats must be eliminated. The ecosystems should not have changed so much, even if the changes are totally natural, that released animals could not survive. Releases work best into top quality habitat, in the core of the historic range. The habitat should be protected and large enough to support a viable population, allowing for an increase in numbers to reach this. Habitat with likely competitors, predators or at the carrying capacity for remaining wild populations of the release species should not be used.

Release stocks
Translocated wild stock generally fares better than captive bred, but wild stock should not be used if this endangers the original population. Captive stock should be trained to cope, either by human handlers or interacting with surrogates or remnant local populations. The released animals should not endanger wild populations by importing disease or genetic and social disruption. The release of large numbers over many years is preferable. There is some evidence that for birds and mammals success rates increase as numbers released rise to 100, but beyond that there is little gain.

Support
Soft releases, providing support in the initial stages with food, shelter and familiarisation work better than abrupt (hard) releases. Techniques for support and monitoring should be properly developed beforehand, combined with knowledge of species' biology to assess chances of success. Sufficient funds should be available not only to set up a release but for long-term support. Releases should accord with local laws, which should be amended to embrace released stock in protective regulations.

Public relations
Public support is vital through education and involvement. A supportive public will be an asset. This may involve incentives or compensation if the released species is perceived as a pest though any releases likely to endanger the public can generally be

discouraged. Relevant government department and NGOs should all be involved with clear lines of responsibility established.

Estimates of the value of release projects vary with criteria used to judge success. A review of releases of threatened animals in Australia, New Zealand, Canada and the USA between 1973 and 1986 estimated only 44 per cent succeeded. A stricter review of releases from 1900 to 1992, anywhere in the world, with criteria of captive bred stock, reaching self-sustaining populations of 500 plus, estimated only 11 per cent success.

Biodiversity banks

Ex-situ conservation uses biodiversity banks, starting with zoos and botanic gardens keeping live animal and plants through to banks of genes, stored seeds and microbes.

Zoos

Zoos began as menageries for status, entertainment and education but over the last 30 years they have consciously adopted a role in conservation, as refuges for the endangered and sources of captive supply for releases. London Zoo has been dubbed the 'Ark in the Park', playing to this theme with advertisements in the early 1990s of endangered animals with the caption 'Without zoos you might as well tell these animals to get stuffed'.

The actual success of zoos has been fiercely contested. The World Conservation Monitoring Centre accepts the 1990 *International Zoo Yearbook* survey listing 878 zoos globally, in 83 countries, with an inevitable bias to the developed world, 298 in Europe versus 16 in Africa. Taxa held are also biased towards charismatic vertebrates. The 878 zoos listed hold 1,232,226 specimens of vertebrates but most were of little importance for conservation. Of 629 mammals that the IUCN recognised as globally threatened the zoos held 20,628 specimens, scarcely 10 per cent of the total of 202,000 mammals held. Only 10 per cent of the zoos were reckoned to make a significant contribution to conservation; 20 full species and a similar number of subspecies have been saved by conservation in zoos. Only nine threatened mammals have captive populations greater than 500. However, the anti-zoo lobby creates a one-sided picture. For example, the critical Zoo Inquiry cited only 405 zoos out of the 10,000 contributing to ISIS, ignoring many species conservation programmes not included under the ISIS scheme. Criticisms of zoos also ignore their greatest impact. Estimates of global visitor numbers to zoos in the 1990 *International Zoo Yearbook* were 619 million, over a tenth of the world's population. The role of zoos both in species survival and education is tremendous. Table 5.3 summarises some zoo successes and problems.

The World Zoo Conservation Strategy (WZCS) 1993 estimated there were 10,000 zoos globally of which 1,000 could take part in conservation work. The Strategy sees

Table 5.3 *Numbers of individuals held in zoos versus numbers actually born in zoos for five charismatic mammals*

Taxon	In captivity	Number born in zoo captivity	Notes
Sumatran rhino, *Dicerorhuns sumatrensis*	13	1	Famous example of much criticised capture programme.
Giant panda, *Ailuropoda melanoleuca*	9	3	The most charismatic mammal of all and subject of furious debate.
African hunting dog, *Lycaon pictus*	269	0	Has bred in captivity outside zoos.
Gorilla, *Gorilla gorilla*	617	300	Successful zoo breeding.
Golden lion tamarin, *Leontopithecus r. rosalia*	592	Most	Much cited success story leading to release.

Source: Groombridge 1992.

zoos as complementary to, not a substitute for, other conservation work. There are five main aims:

1 To identify the areas where zoos can make a contribution to the conservation and sustainable use of natural resources.

2 To develop support and understanding for conservation by zoos from authorities and other social and political organisations.

3 To convince zoo authorities that the greatest purpose they serve is the contribution they can make to conservation.

4 To assist zoos in the formulation of policies including conservation priorities.

5 To augment contributions of individual zoos via contacts within the global zoo and conservation network.

Critics of zoos, led by groups such as Zoo Check, pointed to zoos' discovery of conservation as a ticket-selling exercise, with many of the 1,000 potential zoos poorly managed, the criteria for their inclusion simply membership of a recognised zoo association rather than any real commitment of facilities. The involvement of known bad zoos, transfers of captive breeding stock as part of legitimate projects to poor zoos and zoos as sources of surplus stock to the pet trade and laboratory animal market worsened the problems for good zoos. In Britain the crisis came to a head in 1991. London Zoo, threatening imminent closure due to lack of funds, bought in a Chinese giant panda, Ming Ming, at a time when China was accused of a rent-a-Panda philosophy, depleting wild stocks because captive breeding could not meet demand. Unfortunately Ming Ming flew in when London Zoo's male was in Mexico so a German panda was brought over. To seal the nuptials Cilla Black, hostess of the TV show Blind Date, presided. They did not breed. Soon after the running of London Zoo was restructured in an attempt to focus more on conservation. Zoos are an easy target and their critics missed the point. The potential role of zoos within integrated conservation programmes is vital.

The glare of negative publicity has forced serious consideration of how this can be achieved. Education is a key task. Zoos are largely dependent on visitors for their income but this can be turned into a positive asset, so that visitors realise their entrance fee has also contributed to conservation work as well as learning from their visit. Zoos can concentrate on a limited number of **flagship species**, charismatic or keystone, primarily vertebrates, with the intention to supply release programmes in protected areas. A zoo could redeploy funds and space to build a viable ex-situ population and expertise. The release would protect habitats and other taxa ex-situ. Zoos also remain a last refuge for species reduced to one or two vulnerable wild populations. This integration of in-situ and ex-situ conservation is particularly encouraging but brings its own problems. The sheer number of zoo animals of little conservation value is a burden and cannot be summarily disposed of to free resources and space for target species. London Zoo discovered this in the wake of the changes when rumours of unwanted animals to be put down caused public concern. Zoos in the developed world are vital as training centres, not just for zoo staff but allied conservationists, especially from the developing world. Training can help to transfer skills, and education and above all act as a boost to the self-confidence of staff who often work in thankless isolation. Training should include follow-up contacts to retain the momentum. The most famous example is the International Training Programme in Conservation and Captive Breeding of Endangered Species, based at Gerald Durrell's famed Jersey Wildlife Preservation Trust. This programme is seen as part of integrated conservation, including research, in-situ and ex-situ work. Zoos have increasingly realised that **invertebrate conservation** is viable. Schemes such as the *Partula* snails and British Field Cricket (*Gryllus campestris*) breeding programmes at London Zoo have become national news items. **Aquaria** have an additional role, refining extensive commercial aquaculture

Plate 24 *Jersey Wildlife Preservation Trust training programme. The JWPT International Training Centre works to share and spread skills for conservation, supporting local staff from the developing world. Here a trainee from the island of St Lucia is working with the Curator of Reptiles*

Photograph Tony Allchurch/JWPT.

expertise for conservation and acting as a buffer against the impact of the aquarist fish trade. (See Plate 24.)

Botanic gardens

The first modern botanic garden was founded in Padua in 1545 to collect together the ever-increasing variety of plants brought back by explorers and act as an inventory and resource bank. There are over 1,500 botanic gardens worldwide (inevitably 532 in Europe, 82 in Africa). They harbour 35,000 species, 15 per cent of the world's plant species diversity. The Royal Botanic Gardens, Kew, alone holds 27,000. Some 800 botanic gardens claim to be active in conservation, though problems include underfunding, poor maintenance, taxonomic bias (orchids, succulents and temperate trees are over-represented) and often tiny populations, perhaps just one show specimen. Botanic Gardens Conservation International co-ordinates data and expertise across 327 gardens and many national networks exist. The Botanic Gardens Conservation Secretariat of IUCN co-ordinates priorities for collection and identification of under-represented groups and establishment and training for staff in the developing world. There has been a shift from diverse collections of exotic species to specialisation on taxa, region or plants of particular use. Additional benefits from botanic gardens are historic records such as herbaria, education and links to protected areas. A few species exist solely in botanic gardens, e.g. the Birch tree, *Betula murrayana*, down to one specimen.

Seed, **microbe** and **gene banks** are hi-tech storage facilities. Seed banks are effective for conservation of genetic diversity in many plants. The number of seeds stored represents a viable population taking up very little space. Many can be stored long term, one hundred years or more, and seeds capture the genetic spectrum of a species, compared to a handful of individuals. Storage at low humidity (5–8 per cent) and temperatures (–10 to –20°C) is effective for many seeds. However, non-seeding plants or those producing **recalcitrant** seeds which must germinate rapidly or die cannot be stored. Some 20 per cent of seed plants are recalcitrant, including many tropical and crop species. Even longer lasting **orthodox** seed gradually loses viability and requires growing on to reset seed. Initial set-up costs for storage equipment can be high and vulnerable to breakdown. There were some 528 seed banks linked to botanic gardens in 1990, often specialising by region or taxon. Particular attention has been given to improving banks of wild seed of domestic species and their relatives. Fifty major banks are linked via the Consultative Group on International Agricultural Research and the Centre for Plant Conservation has published guidelines for conservation of endangered plant genetic biodiversity in 1990. The guidelines suggest:

1 Order of priority. Species in danger of extinction, unique taxa, those that can be reintroduced, those with ex-situ potential, those with potential value.

2 Collect seed from five populations per species from throughout their geographical and environmental range.

3 Collect seed from 10–50 individuals per population.

4 The numbers of seeds taken must take account of viability.

5 Collect seeds over several years so as not to impact on local populations.

Field gene banks are an allied technique relying on cultivation of plants in managed sites. Although perhaps within the plant's natural range more intensive ex-situ management is used. Field gene banks exist for many crop and forestry species and are especially useful to maintain local landraces in the conditions which spawned them. **Gene** (or **germplasm**) **banks** store examples, called accessions, of the genetic diversity of species. Thirteen International Agricultural Research Centres have been funded to concentrate on domestic species and their relatives and now hold over 90 per cent of landraces of main crop species, e.g. Centro Internacional de La Papa, Lima, Peru, mandated as a centre for potatoes, with 500 varieties of potato, 1,500 of wild potatoes and 5,200 of sweet potato. Plants have also been conserved in clone banks, using artificial propagation of cuttings and tissues. Seed and gene banks have been the subject of tension over ownership, control and access, especially between source (developing) and recipient (developed) countries.

Microbe banks have grown from co-ordination of existing cultures. The World Federation of Culture Collections accredits 345 cultures from 55 countries which are important as pure sources for research. Their origins have resulted in a bias to disease organisms and a few useful microbes, e.g. nitrogen fixing bacteria. UNEP's promotion of Microbial Resource Centres (16 in 1992) has widened their role to include conservation, collection, training and secure holding facilities.

Conservation of animal genetic diversity has relied less on banks. An FAO/UNEP Animal Genetics Resources Programme was launched in 1982 primarily for domestic breeds, to develop a database and techniques such as cryopreservation of semen, eggs and embryos. A World Watch List of endangered breeds exists which also includes information on such technologies.

Mars calling

August 1996 and as I finish writing this book the news is full of reports that scientists have identified possible evidence of ancient life from Mars. The evidence consists of what might be microfossils of microbial life in a meteorite fragment found in Antarctica (A meteorite? Recovered from the polar ice? Have these scientists never seen that film 'The Thing'?). If the patterns in the rock were once alive the surface of Mars today appears utterly dead. Huge meandering gullies suggest that water once flowed over the surface but no longer. If there was life on Mars it is either long gone or buried deep. Either way the news from Mars makes the richness of life on Earth seem even more precious. It also opens up fresh horizons for biodiversity on an interplanetary scale. We would do well to remember the destructive history of human expansion over this planet before we set off for any others.

Conservation is rife with problems and challenges but this should not eclipse what we do know. The principles and practice of good conservation are well understood throughout the world. The real challenge is whether we can be bothered. Ultimately conservation is not about the details of genetics, zoo management or treaty clauses. Conservation depends on our commitment.

Summary

- Conservation covers species and habitat protection, sustainable use of biodiversity, including ensuring that those doing the conservation reap the benefits plus treaties and funding.

- International and national legislation is used to control biodiversity trade, protected areas and financial aid.

- Protected areas recognise habitat loss as a major cause of degradation. They range from totally protected reserves to multi-use zones.

- Ex-situ conservation includes captive breeding, zoos, botanic gardens and gene banks. Increasingly ex-situ work supports in-situ conservation within protected areas.

Discussion questions

1 If ivory trading can guarantee the survival of the African Elephant should moral concerns still result in an ivory trade ban?

2 What is the strategy for and structure of protected areas in your country?

3 Should commercial companies be allowed to copyright and own natural genetic biodiversity?

See also

Economics for conservation, Chapter 4.
Ecological processes of extinction, Chapter 4.
Origins of Rio Convention on Biodiversity, Chapter 1.

General further reading

'Measures for conservation of biodiversity and sustainable use of its components'. K. Miller, M. H. Allegretti, N. Johnson and B. Jonsson. 1995. In V. H. Heywood (ed.). *Global Biodiversity Assessment*. Section 13. CUP, Cambridge.
Detailed review of approaches to conservation including protected areas, legislation, economics and ex-situ conservation.

Creative Conservation. P. J. S. Olney, G. M. Mace and A. T. C Feistner (eds) 1994. Chapman and Hall, London.
Principles and case studies of captive breeding and reintroduction schemes.

Biodiversity and Conservation. 1995.
Special volume of journal in tribute to Gerald Durrell. Discussion of role of zoos.

'Protected Areas'. J. A. McNeely (ed.) 1994. *Biodiversity and Conservation* (special issue), 3 (5).

Essentials of Conservation Biology. R. B. Primack. 1993. Sinauer Assoicates, Sunderland, Mass.

Principles of Conservation Biology. G. K. Meffe and C. R. Carroll. 1994. Sinuaer Associates, Sunderland, Mass.
Two comprehensive textbooks of exploring conservation policy and practice in the light of biodiversity and changing concepts.

Glossary

aerobic Environment with or metabolic process requiring oxygen.

allele Different forms of a particular gene.

allogenic Biological patterns or processes arising outside the ecosystem which they eventually affect.

allopatric Spatially separate populations. Allopatric speciation occurs when a previously continuous population is split, each subpopulation diverging sufficiently to become different species.

anaerobic Environment lacking or metabolic process occurring in absence of oxygen.

anagenesis New species arising by changes to an existing species through time so that ancestors and descendants are sufficiently different to warrant status as different species. Also known as **chronospeciation** or **phyletic evolution**.

archeabacteria One of two groups of bacteria (the other are Eubacteria) comprising the prokaryotes. Characterised by diverse tolerances to extreme environmental conditions. Broadly equivalent to domain **Archaea**.

assemblage Collection of species living together but without any special associations, mutual dependence and indifferent to precise combination of species present.

augmentation Addition to individuals to reinforce existing population.

autogenic Biological patterns or processes arising within the ecosystem which they eventually affect.

bequest value Valuation of biodiversity based on not using a resource now but estimate of potential worth to future generations.

biocontrol Control of a pest species by use of another organism, e.g. disease, parasite or predator.

biodiversity Totality of genetic, taxonomic, ecosystem and domestic richness and variation within the living world.

biodisparity The degree of difference between components of biodiversity, e.g. biodisparity between a haddock and a snail is greater than that between two species of fish.

biogeography Study of biological patterns and processes with a spatial and often historical context.

biome Global or continental scale ecosystem zones defined by vegetation and fauna largely determined by climate.

bioregion Coherent natural area defined by landscape and species. Contrasts to regions defined by artificial political boundaries.

biosphere The planet's land, sea and air ecosystems characterised by the presence of life.

biosphere people Affluent populations able to use economic power to exploit markets from around the world. Divorced from intimate ties to local environment.

breed A recognised, standard form of a domestic animal species.

CAMPFIRE Communal Areas Management Programme For Indigenous Natural resources, a Zimbabwean scheme to link conservation of wildlife with economic benefit to local people.

CPD (Centre of Plant Diversity) Areas particularly rich in plant life which, if protected, would safeguard the majority of wild plants of the world. Defined by IUCN.

chemoautotrophy Metabolic oxidation of simple inorganic compounds in environment to release energy to power manufacture of organic food compounds.

chimera species Domestic hybrids created from separate species which do not routinely interbreed and retaining features from both parents.

chloroplast Cell organelle that is site of photosynthesis.

chronospeciation New species arising by changes to an existing species through time so that ancestors and descendants are sufficiently different to warrant status as different species. Also known as **anagenesis** or **phyletic evolution**.

cladogenesis Speciation by splitting of ancestral lineage into several surviving lineages, each sufficiently different to warrant status as separate species.

Clovis overkill Losses of north American megafauna between around 12,000 years ago attributed to spread of Clovis hunter culture down through Americas.

community Collection of species living together with implication of strong interdependence and resulting patterns sensitive to precise combination of species present.

contingent valuation Valuation of biodiversity based on amount of money people say they are willing to pay, or forgo, towards conservation.

cross fostering Use of surrogate, usually domestic, species to incubate eggs of threatened species in captivity.

cultivar A recognised, standard form of a domestic plant species.

cyanobacteria Prokaryote allies of true bacteria with ability to photosynthesise, often called blue green algae.

dambo Zimbabwean name for small river valley wetlands found throughout much of southern Africa.

debt-for-nature swap Purchase of part of a country's foreign debt by a conservation organisation who establish conservation projects within the debtor nation in return for waiving debt payments.

deep time The hundreds and thousands of millions of years of geological and evolutionary history.

deoxyribonucleic acid (DNA) The molecule containing the genetic code held within the nucleus of a cell.

domain Classification of life based on analysis of ribosome RNA. Results suggest all life forms belong to one of three domains; Bacteria, Archaea or Eucarya.

EBA (Endemic Bird Area) Sites characterised by high numbers of endemic or restricted range bird species, defined by International Council for Bird Preservation, as part of global analysis of endemicity hot spots.

ecological redundancy The idea that there are more species present in many ecosystems than are required to maintain proper ecosystem function.

ecological refugees Peoples displaced from traditional lifestyle intimately linked to local environment but lacking biosphere people's economic security and often forced to move into new territory.

ecosystem The living, biotic, inhabitants and their physical, abiotic, environment.

ecosystem function The physical outcome of species' activity within an ecosystem typically referring to cycling of chemicals or alteration of the physical environment, e.g. photosynthetic production of oxygen.

ecosystem people Traditional cultures closely wedded to local biodiversity and resources, their economy based on muscle power with rights and responsibilities often maintaining sustainable use of environment.

ecosystem services The benefits to life, including humanity, accruing from some ecosystems functions.

Ediacara Late Pre-Cambrian fauna, named after Ediacara Mountains in Australia where first fossils were found.

endemic A specie found within one country, or sometimes other defined area, e.g. bioregion, and nowhere else.

ethnobotany Traditional folk classifications of plants typically based on medicinal, food and symbolic use.

Eubacteria The modern bacteria plus cyanobacteria, one of two groups of bacteria (the other is **Archeabacteria**) comprising the prokaryotes. Characterised by diverse tolerances to extreme environmental conditions. Broadly equivalent to domain **bacteria**.

Eucarya One of three Domains of life, a classification based on analysis of ribosome RNA suggesting all life forms belong to one of three domains: Bacteria, Archaea or Eucarya.

eukaryote Life characterised by possession of discrete cell organelles, particularly a nucleus, broadly equivalent to domain Eucarya.

ex-situ Conservation of biodiversity away from its natural habitat.

exaptation An adaptation that turns out to be fortuitously beneficial for quite a different purpose to its original use.

existence value Valuation of biodiversity based on value to a person who may never use, see or benefit from it but is willing financially to support its continued existence.

externality Economic cost of using biodiversity not borne by the user but avoided and often passed on so that everyone bears part of the cost, e.g. pollution, which could be treated by an industry at cost to profits but is released degrading everyone's environment.

FAO Food and Agriculture Organisation of the United Nations.

flagship species Charismatic species used as focal point for conservation campaign.

GEF (Global Environmental Facility) Fund set up in 1990 and administered by World Bank to finance protection of the global commons.

gene A discrete, heritable unit of genetic data, consisting of DNA and carrying the code to regulate a particular characteristic.

genotype The genetic constitution of an organism.

head starting Captive rearing and then release of juvenile stages of a species to avoid high mortality of young.

hedonic pricing Valuation of biodiversity based on expert opinion of economic worth.

heterozygosity Genetic variability of individuals and populations of a species.

homozygosity Genetic uniformity of individuals and populations of a species.

inbreeding Reproduction within a small population of related individuals, often reducing fitness.

in-situ Conservation of biodiversity within its natural habitat.

introduction Release of a species into an area outside its natural range.

isotope Forms of the atom of an element varying in numbers of neutron particles and so in mass.

ITTA (International Tropical Timber Agreement) International agreement by cartel of tropical timber producer and consumer countries with recognition of and mechanisms for conservation projects.

IUCN (International Union for Conservation of Nature and Natural Resources) The World Conservation Union.

keystone species Species with a fundamentally important influence on the patterns, processes and functions of their habitat.

kingdom Twentieth-century classification of life forms based on sub-cellular features. This approach suggested that all life can be divided between five kingdoms: Monera, Protista, Animalia, Plantae and Fungi.

K/T boundary The divide between the Cretaceous (standard abbreviation K) and Tertiary (standard abbreviation T) geological periods, famous for mass extinction event that witnessed extinction of dinosaurs.

Lazarus species A species thought extinct but rediscovered.

lichen Symbiotic life form, a sandwich of fungus outer coat sheltering algae.

living dead species A species with one or some individuals still alive but inevitably doomed to extinction as these die out, leaving no progeny.

macrogenesis Speciation by a dramatic mutation, producing significantly different offspring, often termed hopeful monsters.

megafauna Terrestrial animals, primarily mammals, of large body size.

metapopulation A species split into separate populations, linked by dispersal of individuals

microfossil A microscopic fossil, typically used to refer to fossils of microbes.

miner's canary Species or ecosystems showing marked degradation that may be a warning sign of wider environmental damage. Analogy to canaries used by miners as early warning of dangerous gases.

minimum dynamic area The smallest area required to conserve the totality of patterns, processes and functions of an ecosystem.

minimum effective population The smallest number of individuals within a larger population that actually participate in reproduction required to ensure a species' survival into foreseeable future.

minimum viable population The smallest isolated population required to ensure a species' survival into foreseeable future.

mitochondrion Cell organelle that is site for respiration.

nucleotide base Individual molecular building blocks that link to form DNA and RNA, their sequence determining the genetic coding.

OECD Organisation for Economic Cooperation and Development.

option value Valuation of biodiversity based on estimates of potential uses, including insurance against the unknown.

organelle Structures inside Eukaryotic cells specialising in particular task, e.g. mitochondria and chloroplasts.

outbreeding Reproduction between individuals not closely related, typically drawn from a large, heterozygous population.

parapatric speciation Speciation as adjacent populations, stretched over environmental gradients, diverge sufficiently to warrant status as separate species.

peripatric speciation Speciation as small, peripheral subpopulations become isolated from main population and diverge sufficiently to warrant status as separate species.

phenotype The observed characteristics of a species, the result of expression of genotype interacting with environment.

photosynthesis Trapping of light energy to power manufacture of organic food compounds.

photoautotrophy Alternative term for photosynthesis.

phylogeny Evolutionary lineage of a taxon.

Phylum Level of classification below Kingdom, typically based on fundamental body form and reproduction.

polymorphic Variable in form or character.

polyploidy Individuals containing more than one set of genes.

prokaryote Organisms characterised by lack of distinct cell nucleus. Equivalent to the Kingdom Monera or Domains Archaea and Bacteria.

protected area Area with legal designation recognising conservation as sole or one of several purposes of the area.

radiation (evolutionary) Evolutionary diversification, most commonly referring to increase in numbers of taxa.

rebound (evolutionary) Recovery of ecosystem following mass extinction event.

recipient country Country receiving raw biodiversity resource which may then be refined into useful product, e.g. tropical forest plants exported to developed world recipient countries to test for potential pharmaceuticals.

recombination Swapping over of genes between chromosomes at cell division to produce genetic variation in gametes.

Red Data Books and **Red Lists** Compilations of status of and threat to species using internationally agreed categories classifications of risk. International books and lists produced by IUCN have been used as model for national and regional lists.

reintroduction Establishment of new population of a species within historic range from which it has been lost.

RNA, Ribonucleic acid Molecule of nucleic acids, assembled as a copy DNA genetic data base and used to transport genetic data to sites inside cells where proteins are manufactured using the information.

ribosome Site of protein manufacture inside cell, reading the genetic code brought by RNA from original data base of DNA.

rRNA RNA that forms part of structure of ribosomes.

SLOSS (Single Large Or Several Small) Acronym for contentious arguments over best designs for protected areas.

source country Country providing raw biodiversity resource which may then be exported to recipient countries.

speciation Evolutionary formation of new species.

stasipatric speciation One form of sympatric speciation wherein distinctive homozygous subpopulations become sufficiently different to warrant status as separate species.

sympatric Populations or species sharing at least part of their range. Sympatric speciation refers to speciation without any physical separation of ranges.

taxonomy The science of naming and classification of life.

TFAP (Tropical Forestry Action Plan) Programme to halt destruction of tropical forests and promote sustainable use run by FAO.

total economic value The value of biodiversity as the sum total of benefits from use and non-use, including assessment of value from ecosystem function and costs from externalities.

transgenic species Species containing incorporating genes from another species, the result of genetic engineering.

trophic level Position in food chain.

UNEP United Nations Environment Programme.

vicariance Scattered, disjointed distribution of a taxon due to historic dispersal via movement of continents.

WCMC World Conservation Monitoring Centre.

WWF World Wide Fund for Nature (originally the World Wildlife Fund).

zoning Multiple use of a protected area, including conservation but also commercial exploitation, the uses separated between different parts of the area.

zooxanthellae Symbiotic single-celled algae found inside corals and lost during coral bleaching events.

 # Further reading

General

Global Biodiversity. Status of the Earth's Living Resources. B. Groombridge (ed.). 1992. Chapman and Hall, London.

Global Biodiversity Assessment. V. H. Heywood (ed.). 1995. CUP for UNEP, Cambridge. These two volumes are massive reviews, pulling together the entire global expertise in this field. Both cover all the topics in this book (and more) in detail and are rich in additional reading. They are both vital reading (though not all in one sitting).

BioDiversity. E.O. Wilson (ed.). 1988. National Academy Press, Washington DC.
The classic publication that ignited so much work.

The Diversity of Life. E.O. Wilson 1992. Harvard University Press, Harvard.
Beautifully simple and clear exploration of ecology and evolution.

The journal *Biodiversity and Conservation* should also be essential reading. Not only are individual articles relevant but there have been many special editions covering topics such as role of zoos, reintroduction programmes, sustainable use of wildlife.

Specific topics

Origins and concepts

Changes in biological diversity. T.E. Lovejoy. 1980. In *The Global 2000 Report to the President, Vol. 2 (The technical report)* G.O. Barney (ed.). Penguin Books, Harmondsworth. Lovejoy's seminal paper still captures the early themes.

Ecology and living resources biological diversity. In *Environmental Quality 1980: Eleventh Annual Report of the Council on Environmental Quality*, E.A. Norse and R.E. McManus. 1980. Council on Environmental Quality, Washington DC.
The second seminal paper from 1980, with notable inclusions of philosophical topics.

BioDiversity. E.O. Wilson (ed.). 1988. National Academy Press, Washington DC. The classic publication that ignited so much work.

Is humanity suicidal? E.O. Wilson. 1993. *BioSystems*, 31, 235–242. Wilson muses on human destruction of the environment.

What does 'biodiversity' mean – scientific problem or convenient myth? A. Ghilharov. 1996. *Trends in Ecology and Evolution*, 11, 304–306. Biodiversity has become so ubiquitous the concept may be unhelpful.

Preface. *Biodiversity. Measurement and Estimation*. J.L. Harper and D.L. Hawksworth. 1995. Chapman and Hall, London. Reviews origins and rise of the term.

Environmentalism and evolutionary ecology

Green Warriors. D.F. Pearce 1991. Bodley Head, London.

Modern Environmentalism. D. Pepper. 1996. Routledge, London.

A History of Nature Conservation in Britain. D. Evans. 1992. Routledge, London.

The Environmental Sciences. P.J. Bowler. 1992. Fontana Press, London.

The Background of Ecology. P.P. McIntosh. 1985. CUP, Cambridge. Our attitudes to and understanding of nature cannot be divorced from the history of these sciences and the wider social context. These are all readable and revealing explorations into the antecedents of biodiversity.

History of life

Life Pulse. Episodes from the Story of the Fossil Record. N. Eldredge. 1989. Pelican Books, London. Readable summary of life's ups and downs throughout deep time.

Wonderful Life. S.J. Gould. 1991. Penguin Books, London. Inspiring romp through the weird wonders of the early Cambrian and a challenge to our understanding of animal life's subsequent history.

Extinctions in the fossil record. D. Jablonksi. 1995. In *Extinction Rates*. OUP, Oxford. Detailed but readable account of past extinction events.

Bad Genes or Bad Luck? D.M. Raup. 1993. OUP Paperbacks, Oxford. Extinction patterns, causes and lessons from history.

What I did with my research career: or how research on biodiversity yielded data on extinctions. J.J. Sepkoski jr. 1994. In *The Mass Extinction Debates. How Science Works in a Crisis*. Stanford University Press, Stanford. Human and revealing account from one of the originators of recent research into past extinction patterns.

Economics, volcanoes and Phanerozoic revolutions. G.J. Vermeij. 1995. *Paleobiology*, 21, 125–152. An example of the broad links between global systems, evolution and life's history.

Revising our view of life's history

Extraterrestrial cause for the Cretaceous–Tertiary extinction. L.W. Alvarez, W. Alvarez, F. Asaro and H.V. Michel. 1980. *Science*, 208, 1095–1108.
The classic paper suggesting meteor impact as cause of dinosaur extinction.

The Dinosaur Heresies. R. Bakker. 1986. Longman, London.
Right or wrong, Bakker's interpretation of the dynastic struggles to dominate the land challenge the dusty view of life's epic history.

A new metazoan from the Cambrian Burgess shale, British Columbia. S. Conway Morris. 1977. *Palaeontology*, 20, 623–640.
Hallucigenia makes its debut.

New early Cambrian animal and onychophoran affinities of enigmatic metazoans. L. Ramskold and H. Xianguang. 1991. *Nature*, 351, 225–227.
Hallucigenia turns out (or over) to be a worm.

Symbiosis in Cell Evolution: Microbial Communities in the Archean and Proterozoic. L. Margulis. 1993. Freeman, New York.

The current crisis

Chicken Little or Nero's fiddle? A perspective on declining amphibian populations. J.G. Blaustein. 1994. *Herpetologica*, 50, 85–97.
Review of recent amphibian extinctions.

Declining amphibian populations in perspective: natural fluctuations and human impacts. J.H.K. Pechmann and H.M. Wilbur. 1994. *Herpetologica*, 50, 65–84.
Review of recent amphibian extinctions.

Coral bleaching. B.E. Brown and J.C. Ogden. 1993. *Scientific American*, 268, 44–70.
Good general review article.

Coral bleaching in the 1980s and possible connections with global warming. P.W. Glynn. 1991. *Trends in Ecology and Evolution*, 6, 175–178.

Mass extinctions of European fungi. J. Jaenike. 1991. *Trends in Ecology and Evolution*, 6, 174–175.
Less familiar taxa are also suffering extinctions.

Recent animal extinctions: recipes for disaster. D. A. Burney. 1993. *American Scientist*, 81, 530–541.
Evidence of human impacts and the history of the current crisis.

The Miner's Canary. N. Eldredge. 1992. Virgin Books, London.
Highly readable account of current extinctions popularising image of species loss as a warning.

40000 years of extinctions on the 'planet of doom'. P.S. Martin. 1990. *Palaeogeography, Palaeoclimatology, Palaeoecology*, 82, 187–201.

The End of Evolution. P. Ward. 1995. Weidenfeld and Nicolson, London.
The current crisis with a historical perspective through deep time.

Species extinction and habitat loss

Extraterrestrial cause for the Cretaceous–Tertiary extinction. L.W. Alvarez, W. Alvarez, F. Asaro and H.V. Michel. 1980. *Science*, 208, 1095–1108.

Life Pulse. Episodes from the Story of the Fossil Record. N. Eldredge. 1989. Pelican Books, London.

Extinctions and market forces: two case studies. S. Farrow. 1995. *Ecological Economics*, 13, 115–123.
Economics wipes out the passenger pigeon and buffalo herds.

Rarity. K.J. Gaston. 1994. Chapman and Hall, London.
Detailed and insightful coverage of rarity.

The Mass Extinction Debates. How Science Works in a Crisis. W. Glen (ed.). 1994. Stanford University Press, Stanford.
The evidence for and against mass extinctions due to meteor impacts, including discussion of recent advances, e.g. candidate craters.

Human disturbance and natural habitat: a biome level analysis of a global data set. L. Hannah, J.L. Carr and A. Lankerani. 1995. *Biodiversity and Conservation*, 4, 128–155.
Quantification of global terrestrial ecosystem losses.

Extinction Rates. J.H. Lawton and R.M. May (eds). 1995. OUP, Oxford.
Revealing, detailed and thorough collection of articles.

Bad Genes or Bad Luck? D.M. Raup. 1993. OUP Paperbacks, Oxford.
Extinction, historic, recent and causes.

Ecology

Ecological Concepts. The Contribution of Ecology to an Understanding of the Natural World. Cherrett J.M. (ed.). 1989. Blackwell for the BES, Oxford.
Review articles of what (little) we understand.

Large Scale Ecology and Conservation. P.J. Edwards, R.M. May and N.R. Webb (eds). 1994. Blackwell, Oxford.
Many relevant articles tackling problems of scale in ecology and conservation.

Homage to Santa Rosalia, or why are there so many kinds of animals? G.E. Hutchinson. 1959. *American Naturalist*, 93, 145–159.
Santa Rosalia's relics may be the bones of a goat but this is still a classic.

Biological Diversity. The Coexistence of Species on Changing Landscapes. M.A. Huston. 1994. CUP, Cambridge.
Superb synthesis of the ecology of diversity.

Linking Species and Ecosystems. C.G. Jones and J.H. Lawton (eds). 1995. Chapman and Hall, London.
Linking community and ecosystem ecology, with an eye on ecosystem function.

Species Diversity in Space and Time. M.L. Rosenzweig. 1995. CUP, Cambridge. Superb synthesis of the ecology of diversity.
Very different to Huston.

The Diversity of Life. E.O. Wilson. 1992. Harvard University Press, Harvard.
Beautifully simple and clear exploration of ecology and evolution.

Evolution

The Blind Watchmaker. R. Dawkins. 1986. Longman, Harlow.
Awesome exploration of evolution's creative power.

Reinventing Darwin. The Great Evolutionary Debate. N. Eldredge. 1995. Weidenfeld and Nicolson, London.
Evolutionary processes at a larger scale than the mechanics of genes.

The Beak of the Finch. J. Weiner. 1994. Jonathan Cape Ltd, London.
Beautiful tale of research on evolution and speciation among Darwin's Finches.

Analytical Biogeography. A.A. Myers and P.S. Giller. 1988. Chapman and Hall, London.
Excellent textbook, including evolutionary ecology, e.g. speciation.

Evolution: A Biological and Palaeontological Approach. P.W. Skelton. 1993. Addison-Wesley for the Open University, Wokingham.
Comprehensive and well-written textbook.

Ecosystem function and planetary health

Global distribution of natural freshwater wetlands and rice paddies, their net primary productivity, seasonality and possible methane emissions. I. Aselmann and P.J. Crutzen **1989.** *Journal of Atmospheric Chemistry*, 8, 307–358.
Wetlands as source of atmospheric methane. Widely cited estimate of global extent of wetlands.

Coral bleaching in the 1980s and possible connections with global warming. P.W. Glynn. 1991. *Trends in Ecology and Evolution*, 6, 175–178.
Review of coral bleaching linked to global warming.

Linking Species and Ecosystems. C.G. Jones and J.H. Lawton (eds). 1995. Chapman and Hall, London.
Review articles linking species' activities to ecosystem function and health.

Integrating Earth System Science. P. Williamson (ed.). 1995. Special edition, *Ambio*, 23, No. 1.
Review articles of global geochemical cycles, their importance and the politics of wise management.

Taxonomic inventory

Regional patterns of diversity and estimates of global insect species richness. K.J. Gaston and E. Hudson. 1994. *Biodiversity and Conservation*, 3, 493–500.
Good example of recent research into global species total.

Biodiversity. Measurement and Estimation. D.L. Hawksworth (ed.). 1995. Chapman and Hall, London.
Progress, possibilities and problems of inventory work.

Biodiversity, molecular biological domains, symbiosis and kingdom origins. L. Margulis. 1992. *BioSystems,* 27, 39–51.
Fundamental biodiversity – the bacteria in us all.

How many species are there? N.E. Stork. 1993. *Biodiversity and Conservation,* 2, 215–232.
Good example of recent research into global species total.

Wetlands

Wetlands. W.J. Mitsch and T.G. Gosselink.1993. Van Nostrand Reinhold, New York.
Very thorough text including role and function of wetlands.

Global distribution of natural freshwater wetlands and rice paddies, their net primary productivity, seasonality and possible methane emissions. I. Aselmann and P.J. Crutzen. 1989. *Journal of Atmospheric Chemistry,* 8, 307–358.
Global extent of wetlands.

Valuation and management of wetland ecosystems. R. Constanza, S.C. Farber and J. Maxwell.1989. *Ecological Economics,* 1, 335–361.
Overview and examples of role and value of wetlands.

The use of dambos in rural development, with references to Zimbabwe. Dambo Research Unit. 1987. ODA final project report R3869, file ref. ENG 505/512/10 ODA, London.

British Plant Communities. Vol. 4. Aquatic Communities, Swamps and Tall Herb Fens. J.S. Rodwell (ed.). 1995. CUP, Cambridge.
British National Vegetation Classification for many wetland types.

Developing wetland inventories in southern Africa: a review. A.R.D. Taylor, G.W. Howard and G.W. Begg. 1995. *Vegetatio,* 118, 57–79.
Overview of wetlands in southern Africa, including Zimbabwe.

Conservation concepts

Policy for Wild Life. Department of National Parks and Wild Life Management. 1992. NPWLM, Harare.
'Wildlife is a unique economic resource.'

The role of local people in the successful maintenance of protected areas in Madagascar. J.C. Durbin and J.A. Ralambo. 1994. *Environmental Conservation,* 21, 115–120.
Lovely example of problems and possibilities for involving local people in conservation.

Sharing the Land. Wildlife, People and Development in Africa. K. Makumbe (ed.). 1995. IUCN, Cambridge.
Short but comprehensive overview of shifting focus of conservation.

Ecology and conservation; our Pilgrim's Progress. In M. Nicholson *Conservation in Progress.* 1993. John Wiley and Sons, Chichester.
A founding father of post-war conservation muses on recent developments.

Evolutionary ecology and the conservation of biodiversity. J.N. Thompson. 1996. *Trends in Ecology and Evolution*, 11, 300–303.
A discussion of conservation and biodiversity in context of lessons from ecology and evolution.

Conservation practice

Essentials of Conservation Biology. R.B. Primack. (1993). Sinauer Associates, Sunderland, Mass.

Principles of Conservation Biology. G.K. Meffe and C.R. Carroll. 1994. Sinauer Associates, Sunderland, Mass.

Fundamentals of Conservation Biology. M.L. Hunter. 1996. Blackwell, Oxford.
Three from a new wave of very thorough, detailed textbooks reviewing conservation in the light of biodiversity's conceptual rise.

Conservation Genetics. V. Loeschcke, J. Tomink and S.K. Jain. 1994. Birkhauser Verlag, Boston.
Advanced text but relevant and worth the effort.

Protected Areas. J.J. McNeely. 1993. *Biodiversity and Conservation*, special issue, 3, No. 5.

Economics for conservation

Wildlife in the Market Place. T.L. Anderson and P.J. Hill (eds). 1995. Rowan and Littlefield, Lanham.
Examples and background to sustainable use of wildlife as a conservation policy.

Dealing in diversity. V.M. Edwards 1995. *America's Market for Nature Conservation*. CUP, Cambridge.
You can now shoot African big game on ranches in America. Here's why.

The Zimbabwean Communal Areas Management Programme for Indigenous Resources (CAMPFIRE). S. Metcalfe. 1993. Zimbabwe Trust, Harare.
Zimbabwean policy review.

Ecotourism, biodiversity and local development. H. Goodwin and I.R. Swingland (eds). 1996. *Biodiversity and Conservation*, 5, No.3.
Collection of papers reviewing background and examples of ecotourism and conservation.

Wildlife species for sustainable food production. J.E. Cooper (ed.). 1995. *Biodiversity and Conservation*, 4, No. 3.
Collection of papers reviewing sustainable use of wildlife as conservation tool.

Economics of biodiversity

The Economic Value of Biodiversity. D. Pearce and D. Moran. Earthscan, London.
Good introductory textbook to topic including data on wetlands.

Valuation and management of wetland ecosystems. R. Constanza, S.C. Farber and J. Maxwell. 1989. *Ecological Economics*, 1, 335–361.
The role and value of wetlands.

Economics and Biological Diversity. J.A. McNeely. 1988. IUCN, Gland.
Good introductory textbook to topic.

Biodiversity Loss. Economic and Ecological Issues. C. Perrings, K-G. Maler, C. Folke, C.S. Holling and B-O. Jansson. 1995. CUP, Cambridge.
Good introductory textbook to topic.

Biodiversity Prospecting: Using Genetic Resources for Sustainable Development. W.V. Reid, A. Sittenfeld, S.A. Laird, D.H. Janzen, C.A. Meyer, M.A. Gollin, R. Gamaz and C. Juma 1993. World Resources Institute, Washington D.C.

Paradise Lost? The Ecological Economics of Biodiversity. E.B. Barbier. 1994.
Detailed and generally well combined in a thoughtful textbook.

Ex-situ conservation

Parks or arks?: where to conserve threatened mammals? A. Balmford, N. Leader-Williams and M.J.B. Green. 1995. *Biodiversity and Conservation*, 4, 595–607.
The economics and efficacy of conservation in captivity versus the wild.

The release of captive bred snails (*Partula taeniata*) into a semi-natural environment.
P. Pearce-Kelly, G.M. Mace and D. Clarke. 1995. *Biodiversity and Conservation*, 4, 645–663.
Partula turgida may be no more but other *Partula* snails are the focus of fine work.

World Zoo Conservation Strategy: A Blueprint for Zoo Development. R. Wheater. 1995. *Biodiversity and Conservation*, 4, 544–522.
Zoos' future role.

Bibliography

Barbier E.B., Burgess J.S. and Folke C. 1994. *Paradise Lost? The Ecological Economics of Biodiversity*. Earthscan, London.

Bells S., McCoy E.D. and Mushinsky H.R. (eds) 1991. *Habitat Structure*. Chapman and Hall, London.

Bibby C.J. (ed.) 1992. *Putting Biodiversity on the Map*. International Council for Bird Preservation, Cambridge.

Bowler P.J. 1992. *The Environmental Sciences*. Fontana Press, London.

Collinson N.H., Biggs J., Corfield A., Hodson M.J., Walker D., Whitfield S.M., Williams P.J. 1995. Temporary and permanent ponds – an assessment of drying out on the conservation value of aquatic macroinvertebrate communities. *Biological Conservation*, 74, 125–133.

Cooper J.E. (ed.) 1995. Wildlife species for sustainable food production. *Biodiversity and Conservation*, 4, No. 3.

Cotinga (1996) Lears Macaw: a second population confirmed. *Cotinga*, February 96, 10.

Craig P., Trail P., Morrell T.E. 1994. The decline of fruit bats in American Samoa due to hurricanes and overhunting. *Biological Conservation*, 69, 261–266.

Dawkins R. 1986. *The Blind Watchmaker*. Longman, Harlow.

Edwards V.M. 1995. *Dealing in Diversity. America's Market for Nature Conservation*. CUP, Cambridge.

Eldredge N. 1992. *The Miner's Canary*. Virgin Books, London.

—— 1992. *Systematics, Ecology and the Biodiversity Crisis*. Columbia University Press, New York.

—— 1995. *Reinventing Darwin. The Great Evolutionary Debate*. Weidenfeld and Nicolson, London.

Farrow S. 1995. Extinctions and market forces: two case studies. *Ecological Economics*, 13, 115–123.

Flint M. 1991. *Biodiversity and Developing Countries*. ODA, London.

Gaston K.J. 1994. *Rarity*. Chapman and Hall, London.

Gehrels T. 1996. Collisions with comets and asteroids. *Scientific American*, 274, 34–39.

Gilpin M.E. and Hanski I. (eds) 1991. *Metapopulation Dynamics*. Harcourt Brace Jovanovich, San Diego.

Glen W. (ed.) 1994. *The Mass Extinction Debates. How Science Works in a Crisis*. Stanford University Press, Stanford.

Goldsmith F.B. and Warren A. (eds) 1993. *Conservation in Progress*. John Wiley and Sons, Chichester.

Goodwin H. and Swingland I.R. (eds) 1996. Ecotourism, biodiversity and local development. *Biodiversity and conservation*, 5, No. 3.

Gould S.J. 1989. *Wonderful Life*. Hutchinson, London.

Groombridge B. (ed.) 1992. *Global Biodiversity. Status of the Earth's Living Resources*. Chapman and Hall, London.

Hairston N.G., Smith, F.E. and Stobodkin, L.B. 1960. Community Structure, population control and competition. *American Naturalist*, 94, 421–425.

Hannah L., Carr J.L. and Lankerani A. 1995. *Biodiversity and Conservation*, 4, 128–155.

Hawksworth D.L. (ed.) 1995. *Biodiversity. Measurement and Estimation*. Chapman and Hall, London.

Heywood V.H. (ed.) 1995. *Global Biodiversity Assessment*. CUP for UNEP, Cambridge.

HMSO. 1994. *Biodiversity. The UK Action Plan*. HMSO, London.

Hunter M.L. 1996. *Fundamentals of Conservation Biology*. Blackwell, Oxford.

Jablonksi D. 1995. Extinctions in the fossil record. In *Extinction Rates*. OUP, Oxford.

Juste J., Fa J.E., Delval J.P. Castroviejo J. 1995. Market dynamics of bushmeat species in Equatorial Guinea. *Journal of Applied Ecology*, 32, 454–467.

Kreuter U.P and Workman J.P. 1994. Cost of overstocking on cattle and wildlife ranches in Zimbabwe. *Ecological Economics*, 11, 237–248.

Lawton J.H., Nee S., Letcher A.J. and Harvey P.H. 1994. Animal distributions: patterns and process. In *Large Scale Ecology and Conservation*. Blackwell, Oxford.

Lovejoy T.E. 1980. Changes in Biological Diversity. In *The Global 2000 Report to the President, Vol 2 (The Technical Report)*. G.O. Barney (ed.). Penguin Books, Harmondsworth.

MacArthur R.H. 1972. *Geographical Ecology. Patterns in the Distribution of Species*. Harper and Row, New York.

McCarthy M.A., Franklin D.C., and Burgman M.A. 1994. The importance of demographic uncertainty – an example from the Helmeted Honeyeater *Lichenostomus melanops cassidix*. *Biological Conservation*, 67, 135–142.

McNeely J.A. 1988. *Economics and Biological Diversity*. IUCN, Gland.

Madsen T., Stille B., Shine R. 1996. Inbreeding depression in an isolated population of adders, *vipera berus. Biological Conservation*, 75, 113–118.

Margulis L. 1993. *Symbiosis in Cell Evolution: Microbial Communities in the Archean and Proterozoic*. Freeman, New York

Margulis L. and Sagan D. 1987. *Microcosmos. Four Billion years of Microbial Evolution*. Allen and Unwin, London.

May R.M. 1989. Levels of organisation in ecology. In *Ecological Concepts*. Blackwell for the BES, Oxford.

Meffe G.K. and Carroll C.R. 1994. *Principles of Conservation Biology*. Sinuaer Associates, Sunderland, Mass.

Murdoch W.W. 1966. 'Community structure, population control and competition' – A critique. *American Naturalist*, 100, 219–226.

Myers A.A. and Giller P.S. 1988. *Analytical Biogeography*. Chapman and Hall, London.

National Science Board. 1989. *Loss of Biological Diversity: A Global Crisis Requiring International Solutions*. National Science Board, Washington, DC.

Nicolaon K.C., Guy R.K. and Potier P. 1996. Taxoids: new weapons against cancer. *Scientific American*, June 1996, 84–88.

Norse E.A. (ed.) 1993. *Global Marine Biological Diversity*. Island Press, Washington DC.

Norse E.A. and McManus R.E. 1980. Ecology and living resources biological diversity. In *Environmental Quality 1980: Eleventh Annual Report of the Council on Environmental Quality*. Council on Environmental Quality, Washington DC.

Norse E.A., Rosenbaum K.L., Wilcove D.S., Wilcox B.A., Romme W.H., Johnston D.W. and Stout M.L. 1986. *Conserving Biological Diversity in our National Forests*. The Wilderness Society, Washington DC.

Pearce D. and Moran D. 1994. *The Economic Value of Biodiversity*. Earthscan, London.

Pepper D. 1996. *Modern Environmentalism*. Routledge, London.

Perrings C., Maler K.-G., Folke C., Holling C.S. and Jansson B.-O. 1995. *Biodiversity Loss. Economic and Ecological Issues*. CUP, Cambridge.

Postgate J. 1994. *The Outer Reaches of Life*. CUP, Cambridge.

Primack R.B. 1993. *Essentials of Conservation Biology*. Sinauer Associates, Sunderland, Mass.

Redford K.H. and Stearman A.M. 1980. Forest dwelling native Amazonians and the conservation of biodiversity: interests in common or in collision? *Conservation Biology*, 7, 248–255.

Reid W.V., Sittenfeld A., Laird S.A., Janzen D.H., Meyer C.A., Gollin M.A., Gamaz R. and Juma C. 1993. *Biodiversity Prospecting: Using Genetic Resources for Sustainable Development*. World Resources Institute, Washington DC.

Robinson J.G. 1993. The limits to caring: sustainable living and the loss of biodiversity. *Conservation Biology*, 7, 20–28.

Rosenzweig M.L. 1995. *Species Diversity in Space and Time*. CUP, Cambridge.

Stone C.P. and Stone D.B. (eds) 1989 *Conservation Biology in Hawai'i*. University of Hawai'i, Honolulu.

Udvardy M.D.F. 1975. *A Classification of the Biogeographical Provinces of the World*. IUCN, Gland.

Walgate R. 1990. *Miracle or Menace? Biotechnology and the Third World*. Panos Publications, London.

Ward P. 1995. *The End of Evolution*. Weidenfeld and Nicolson, London

Wheatley N. 1994. *Where to Watch Birds in South America*. Christopher Helm, London.

Wilson E.O. 1988. *BioDiversity*. National Academy Press, Washington DC.

—— 1992. *The Diversity of Life*. Harvard University Press, Harvard.

Witt S.C. 1985. *Biotechnology and Genetic Diversity*. California Agricultural Lands Project, San Francisco.

World Resources Institute. 1992. *Global Biodiversity Strategy*. WRI, Washington DC.

Index

Note: page numbers in *italics* refer to tables, figures or illustrations where these are separated from the text.